Are You on Target?

According to the American Heart Association, most healthy Americans should have an exercise target heart rate ranging from 50%-75% of their maximum heart rate. To determine your maximum heart rate, subtract your age from the number 220. Next, determine your heart's beats per minute (BPM) during an exercise routine. To do this, count your pulse rate for ten seconds, then multiply the number of pulses by six. If this number falls in the range of 50%-75% of your maximum heart rate, you're right on target. If it exceeds this range, we encourage you to visit your doctor.

BODYSTORIES

BODYSTORIES

A GUIDE TO EXPERIENTIAL ANATOMY

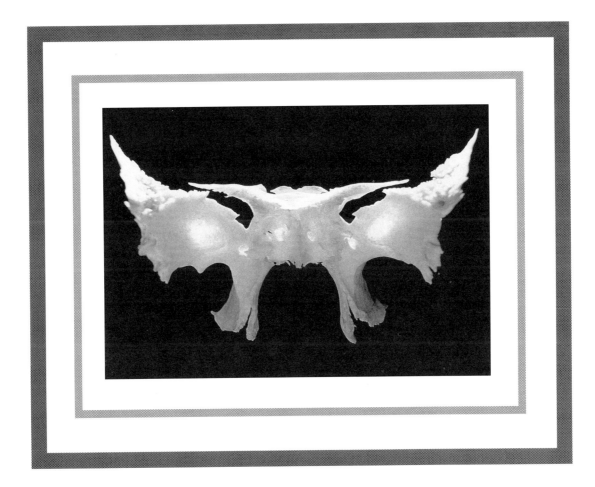

ANDREA OLSEN

IN COLLABORATION WITH CARYN MCHOSE

❖

STATION HILL PRESS

BARRYTOWN, NEW YORK

Published by Station Hill Press, Inc., Barrytown, New York 12507.

Produced by the Institute for Publishing Arts, a not-for-profit, tax-exempt organization in Barrytown, New York, which is supported in part by grants from the National Endowment for the Arts, a federal agency in Washington, D.C., and the New York State Council of the Arts.

Book design by Karen Murley, Full Circle Desktop Publishing & Design, Cerro Gordo, Illinois. Produced on a Macintosh II using Quark Express and Illustrator 88, Palatino type face.
Title page photo: Sphenoid bone of the skull, by Erik Borg.

Library of Congress Cataloging-in-Publication Data

Olsen, Andrea.
    Bodystories: a guide to experiential anatomy / Andrea Olsen, in collaboration with Caryn McHose.
        p.   cm.
    Includes bibliographical references and index.
    ISBN 0-88268-106-0
    1. Human anatomy. 2. Human anatomy -- Problems, exercises, etc. 3. Experiential learning. I. McHose, Caryn.   II. Title. III. Title: Body stories.
    QM23.2.038 1991   91-26293
    611--dc20   CIP

Manufactured in the United States of America.

I would like to acknowledge the work of the following
individuals and their influence on my life:

Janet Adler,
whose work in Authentic Movement brings an understanding
of the body and the psyche.

Bonnie Bainbridge Cohen,
whose experiential teachings of the body systems and
human developmental and reflex patterns
integrate the mind and the body.

Caryn McHose,
whose intuitive methods and progression
for teaching human anatomy form the core of this text.

Gordon Thorne,
whose belief in the value of empty space
provides time and place for creative work.

John M. Wilson,
whose articulate concern for training theatre performers
combines arthrometric analysis with humane philosophy.

❖

This book is dedicated to
Alison
who reminded me
that the function of a book
about anatomy is
not to demystify the body –
it is to help
embody
the mystery.

BODYSTORIES

A GUIDE

TO EXPERIENTIAL ANATOMY

Visual Images

I have chosen the work of twelve artists in the areas of painting, photography, sculpture, furniture making, architecture, and costume design to accompany the text. Their work tells its own story in relation to the body, and asks participation in making the connection between image and word. Multicultural images are included for a world perspective. Movement sketches, anatomical illustrations, and photographs are used to clarify experiential work. Children's drawings enliven the writing with their own view of the body.

Andrea Olsen and Caryn McHose

"IMPECCABLE IN THE BEGINNING"

When Janet Adler would begin a workshop on Authentic Movement, she would call the group together and speak about the importance of beginnings. She reminded us to ask our questions and prepare ourselves, so that we could be fully present in the work. The expression she used was to be "impeccable in the beginning;" to acknowledge what we need to do in order to begin. For me, beginning this text necessitated writing about my origins.

THE TEXT:
Origins

My first course related to anatomy was high school Zoology where we learned the interrelatedness of the species and experienced the unveiling, through dissection, of evolutionary history. College brought Introduction to Biology, where the cell and fantasies of life under the microscope took prominence. Two years later, I had my first course about the human body. I was an undergraduate majoring in art, and was teaching dance at Millikin University, a liberal arts school in the Midwest. The course was taught in the Physical Education Department and the football team and I learned about the muscles. I remember staring at the pictures, memorizing the names, and wondering how they all worked together. My next course was Anatomy and Kinesiology for Dance, taught in graduate school at the U. of Utah by Dr. John M. Wilson. This brilliant course demonstrated in content and teaching style that every science is a philosophy. Dr. Wilson incorporated the principles of Margaret H'Doubler, pioneer in combining anatomy and dance in an educational setting, and spoke of the multidimensionality of the human species. I was captivated by every word and repeated the course twice more in my years of association with Dr. Wilson.

After graduate school I toured with a modern dance company and applied what I had learned to the experience of performing. I taught Anatomy for Dance in workshop settings and developed an approach to dance technique based on anatomical principles. During this time, I became Director of Dance at Mount Holyoke College and taught my first anatomy course. I had bright students, eager with questions, who were obviously as inspired as I had been by learning about the body. I invited Bonnie Bainbridge Cohen, founder of the School for Body-Mind Centering, to speak to the class as a guest lecturer. Bonnie had trained as an occupational therapist; she had also studied dance with Eric Hawkins. She had notable success working with patients, affecting change through her hands, but she couldn't articulate how or why the rehabilitation occurred. Dissatisfied with the constraints of that field, she traveled to England to study with Dr. Karl and Berta Bobath and work with brain disabled children. She also studied Katsugen undu, life-force movement, in Japan with Haruchi Noguchi. This is a method of allowing internal movement to be expressed outwardly, by allowing automatic movement to emerge. Then she and her husband Len, a chiropractor and student of aikido, returned to New York to co-found The School for Body-Mind Centering. With a group of students, they worked experientially, through movement and touch and body listening, to explore what was "not known" about the body. They searched contemporary research and historical texts to support their often unusual findings, and developed principles for teaching the work. Bonnie invited me to join her workshops and classes and to offer my experience as a dancer and choreographer to her own resources.

Writing

When I was about five years old, I remember standing at my child-size table, knowing I should write a book. I was concerned that once I started to read, I would forget what I knew. I didn't write the book, but I did devise an elaborate system of reading where I would memorize without really absorbing the information.

❖

During graduate school, I was involved in a paper for a dance philosophy course. After staying up all night, pondering certain ideas, I raced into the office of my professor and said, "I got it, I understand what we've been talking about." "Good," he said. "I look forward to reading your paper." I responded in complete surprise, "But I don't have a paper. I understand it." "The point is," he said with a smile, "to communicate your understanding to me."

❖

A friend and I were walking down the street. I said, "You really only hear what you are ready to learn." She replied, "You also only hear what someone is willing to tell you."

5

At this time I also met Janet Adler, a movement therapist and founder of the Mary Starks Whitehouse Institute. Her focus was Authentic Movement, a body-oriented therapy with which she had worked for many years. She gathered a group of students to study and eventually articulate the relationship between the witness and the mover in this form. She was also interested in exploring Authentic Movement as a resource for choreography, allowing the psychically charged movement she witnessed in the therapeutic setting to evolve within the context of creative work. The task was to bring the movement's expressive nature to the stage with consciousness, without violating the timing necessary for development and integration. I have worked with these two exceptional colleagues in the study of the body – its science, psyche, and creative potential – for the past ten years.

Many of our explorations took place in a beautiful studio provided by Gordon and Anne Thorne. I had left the college environment to tour more extensively and to explore; and both Gordon, a painter, and Anne, a teacher of creative work with children, became vital influences. We shared a focus on the creative process as our avenue for developing the whole person, and in this sense, to healing. And we collaborated on projects knowing that the images, visions, and interactions that emerged would guide us if we gave them our attention. Gordon contributed to the artistic vitality of the community, both by his painting, and by maintaining an empty space for exploration of process and presentation of work. In spite of people's urging to install permanent fixtures and walls, Gordon's vision was that the space be essentially empty, and that it cyclically return to neutral for new creative work to emerge. Gordon taught me to listen to space as I listen to my body – to value the natural state as equal to what we might do to it.

Both my mind and my body were so filled with influences at this point, that I retreated to work alone. I began to choreograph and to perform as a soloist, and moved to Middlebury College to become Director of Dance and Artistic Director of the Dance Company of Middlebury. In this setting, I met Caryn McHose who was teaching Anatomy and Kinesiology in the dance program. Caryn was a self-taught anatomy teacher, trained since childhood in Dance Improvisation with Betty Jane Dittmar, and held a degree in painting. I observed her classes and watched with amazement as undergraduates from all disciplines and interests walked into the studio, lay down on the floor with eyes closed, and began to concentrate on their bodies experientially. This course was a revelation to me: students hunger for information and experience of the body. Caryn had moved to Vermont, and in her words, lay on the floor of her cabin every day for two years and taught herself anatomy, accompanied by Mabel Todd's book *The Thinking Body*. Caryn's students, in the process of a twelve week semester, did the same thing, transforming their bodies into models of efficient alignment. The effectiveness of her teaching was that she only taught what she had experienced herself. I took Caryn's class for three semesters, and then began teaching a second section of this very popular course. My own approach integrated the analytical, philosophical work of Dr. Wilson, and the principles and touch and repatterning techniques of Bonnie Bainbridge Cohen, with Caryn's direct method of presenting material so that it is immanently understandable at the body level. That course, after three years

Chair: Kristina Madsen

of refining, is the basis for this text. We teach what we need to learn, I am told. As I complete this phase of my own study, the text is offered as a synthesis of my learning experiences for use as is appropriate to you. The questions which arise from the reading and doing become yours to explore.

BODY LISTENING

Storytellers of the Ashanti tribe in Africa begin by saying, "I am going to tell you a story. It is a lie. But not everything in it is false."As I write these words and tell my stories, I am reminded that no description holds the truth. Words can point towards an experience, but they cannot replace it. This is an experiential anatomy text and various approaches to learning are included: factual information, personal stories, evocative and descriptive images, and guided movement explorations. Although each "day" or learning session is presented in a format of one-hour time spans, with time for integration before going on, the words and descriptions must be activated by you. The overview of thirty-one days can be perceived as a month of one-hour sessions, or a twelve-week course meeting three times a week, or a progression to do at your own pace. As movement pioneer Margaret H'Doubler says, in working with the body "you are your own textbook, laboratory and teacher."* The body is our guide, all we need to do is to learn to listen.

PREPARATION FOR USING THE TEXT

How do you learn? Do you need to read, to write, to move, to draw, to touch, to question, to be told, to tell, to be encouraged, to compete, to be left alone or to work in dialogue with someone else? One of the intentions of this book is to let your process become clearer; then you can facilitate your learning. Find a space that you consider private where you can work alone. A wood floor is preferable, and the space should be warm and comfortable. Wear loose-fitting clothing; no shoes or socks. Have pencils and a journal for your own notes and drawings. Establish a realistic schedule for work. Consistency of space and time will be helpful in developing a dialogue with your body.

WAYS OF WORKING: "TO DO" SECTIONS

On your own: Read (or tape record yourself reading) the "to do" sections first, then do. As you become familiar with working in the body, it will be easy to follow the words. With a partner, or in a group: Have one person read aloud as you work; change roles. Use the margins in the book to record your experiences.

COMMON HERITAGE

As we move into a world culture, celebrating the differences in race, nationality, and religion, our awareness of our bodies as our common heritage is increasingly important. There is no age or place, I think, where knowledge of the body is without use. My mother taught first grade for twenty years and says that when she could teach a child to skip, she could teach them to read. Now she teaches swimnastics to her fellow senior citizens, using the principles of this book, with tremendous results in mobility (and stability). Let's consider the text as a map, and enjoy the journey.

*See Margaret H'Doubler's *Dance; A Creative Art Experience.*

Colors

A visual artist talked about her years of working with the human figure: "The body is like a painting. Every time I look, I see something new. When I started studying anatomy years ago as an art student, I wanted to know the structural details, how the parts fit together, and the names of everything. It seemed a good place to start. But then I got involved in the relationship between the mind and the body, and I wanted to explore the motivations behind the forms that I saw. Later, I was drawn to experiential work. I was ready to explore what it felt like to be inside the body rather than looking at it. Always, the body surprised me as a source for discovery. So," she grinned, "the body is full of colors. You can look at each one individually, or step back and experience the whole."

BODYWORK: SOURCES AND RESOURCES

Bodywork is a term used to refer generally to therapeutic techniques of working with the body. The twentieth century has seen a rich development of this work. In the 1930's – 70's individuals such as Mabel Todd, Lulu Swiegard, Ida Rolf, Nikolais Alexander, Moshe Feldenkrais, Milton Trager and others pioneered research in the field and published writings about their findings. These individuals began with their own experience of the body and explored their predilections intellectually and physically. They necessarily had to limit their point of view to an individual perspective for depth and clarity. For example, Mabel Todd in her classic book, *The Thinking Body, A Study of the Balancing Forces of Dynamic Man,* published in 1937, focused on the skeleton and structural alignment. Lulu Swiegard's involvement with the effects of the nervous system on body alignment through imagery and visualization is recorded in her book, *Human Movement Potential, Its Ideokinetic Facilitation,* published in 1974. Ida Rolf in her text *Rolfing, The Integration of Human Structures,* 1977, focused on the fascia and on the integration of body and psyche. Each system was unique in itself, and a student of bodywork was required to read about each method and its innovator to ascertain the common principles. Currently, ancient traditions of body work from diverse cultural origins are becoming available to the public both through translations of writings and through hands-on practitioners. As individuals blend old forms or pioneer new work, texts have appeared which give an overview of the principles common to the field. Deane Juhan's text, *Job's Body, A Handbook for Bodywork,* provides a source rich in scientific fact and experiential principles. The insightful writings of Bonnie Bainbridge Cohen, co-director of The School for Body-Mind Centering, are available through *The Contact Quarterly, A Journal for Moving Ideas,* and are especially illuminating concerning early developmental patterns and reflexes.

Experiential Anatomy has developed parallel to bodywork with a focus on body education. It encourages the individual to integrate information with experience. Thus, experiential anatomy enhances bodywork by providing an underlying awareness of body structure and function. *The Anatomy Coloring Book,* written by Wynn Kapit and Lawrence Elson is a must for both the serious and casual student of anatomy, and David Gorman's hand-drawn, three-volume text, *The Body Moveable,* is a rich source for visual images and text focusing on movement. Gerard Tortora and Nicholas Anagnostakos's *Principles of Anatomy and Physiology* provides an excellent resource for detailed study. These are a few of the many books and articles listed in the bibliography which I have found valuable for the interested reader. In general, what draws your attention is of interest to you. In working with your body, you are the expert. ❖

BASIC CONCEPTS:
Change, Posture, Structure, Choice

*"Who are **you**," said the Caterpillar . . . . Alice replied rather shyly, " I - I hardly know, Sir, just at present – at least I know who I **was** when I got up this morning, but I think I must have been changed several times since then."*

Lewis Carroll, ***Alice's Adventures in Wonderland***

Our bodies are dynamic entities. Our cells are reproducing, processing, and dying constantly as we live. Within a year, a month, the time it takes to read these words, we literally are not the same person we were before. **Change** is constant throughout the life cycle of the body.

Structure is our physical body: the bones, muscles, and other tissues which comprise our bodies. Structure is affected by our heredity and by our life experience in terms of nutrition, illness, and body use and abuse. **Posture** is the way we live in our structure – the energy and attitudes which moment by moment shape our bodies. Our posture affects our structure, and our structure affects our posture, and both can change. For example, if we are born with an extra vertebra or curved lower legs, our posture will be affected by our structure. If we stand with our head forward for many years, our bones will respond to the stresses of our posture. We can observe this dialogue between posture and structure by looking around a room at a group of people: we can see that we share a common structure, but the way we inhabit that structure is very different.

Both posture and structure are about **choice**. We choose how we live in our bodies and our life choices affect our underlying structure. A healthy body remains able to respond – **responsible** – to the changes in situations, people, and personal growth which occur moment by moment throughout our lives. ❖

Claiming Your Height

I lived one summer in the house of a bright, young anesthesiologist who was involved in heart transplant surgery. He was also a runner and complained to me of back problems. We worked with his alignment and noticed that his chest was retreated and that this, combined with a forward head, put stress on his lower back. As we brought his posture into vertical alignment, he took a deep breath and said, "I could never stand like this. I would threaten my colleagues and my patients." He had unconsciously adopted a posture that was nonthreatening and noncompetitive in order to work in an environment which was both. It had given him a certain amount of emotional safety while he developed in his career, but it was now literally hurting him. The question became, was he ready to stand at his full height?

Drawing your skeleton
15 minutes

✎ Draw your skeleton. Rely on what you know and remember, and what you can feel by touching and imagining body parts as you work.

✎ Draw a view from the front, and one from the side. Be as detailed as possible. (No checking pictures.)

ATTITUDES
ABOUT THE BODY

The lack of information about the human body in our years of education is startling since it is our home for our entire lifetime. It seems we either think that the body is too simple and too "physical" to warrant attention, or that it is so complex that it is reserved for medical students. In fact, it is both. It is very simple, and everyone can understand body principles and learn the names of bones, muscles, and organs. It is also the most complex living form. The study of the human body involves both mystery and fact: there is much that is known and equally as much that is left unknown. This paradox suggests that we need to value both the information and the questions about what it means to be human.

One of the most thoroughly neglected areas of body education is the awareness of what is happening inside: the dialogue between inner and outer experience in relation to the whole person. We spend much of our time involved in outer perception through the specialized sense organs of sight, sound, taste, smell, and touch. We are generally less involved in developing our capacities for inner sensing which is the ability of the nervous system to monitor inner states of the body. How and why do we progressively close down our capacity for body listening? As children we are necessarily involved in our relationship to the outer environment for survival. An early aspect of body awareness is about control. One is supposed to gain control over the body as soon as possible to avoid doing

Running

My strongest movement memory as a child is running through the fields in Illinois. When I think of all my years of dance training, I know that this sensation of running, without boundaries, in the flatlands of the midwest is basic to my love of movement.

❖

One of my favorite aspects about living in Utah, was going out at the end of the day into the sunset, and facing the long, downwards hill to my apartment. With a slight shift of weight, I would fall and be running, exhilarated, through space and time.

❖

At a dance workshop, a choreographer asked us to run. She was attempting to clarify performance intention. "Are you running away from, or towards," she asked. I am just running, I thought. The question puzzled me.

Painting: Robert Ferris
"November Light"

❖

Because I travel a lot, flights became a new form of running. Flight reservations often involved considerable mental confusion, as their name suggests. Sometimes I would be completely calm. But often I would get in a state of panic. I would flip from mild depression, unmotivated to go anywhere or do anything, to creating gigantic plans and making twenty decisions, many of which would be changed. My friends called it my "flight pattern." It helped to name the state of confusion; it seemed to make it more tangible, more humorous.

❖

One summer I was making flight plans for California. I felt the familiar flutter. I was writing about emotions and began thinking about the connections. A friend said, "Depression is repressed feeling." As I thought about this, I saw that panic, for me, was a response to numbness. The flutter created stimulation, even though it was confusing. Feeling bad was better than not feeling. "When you are trying to travel, let yourself feel what you're leaving, and where you are going. Recognize the emotion under depression or panic: the pain, fear, and joy around coming and going. Look for what isn't being expressed." And I thought about the connection of running to flying and dancing: they suspend time; they remove me from the real world of emotions, responsibilities, and interactions; and they are experienced through the body. They can be used to go towards or to go away from awareness.

anything embarrassing or terrible in a social context. After control comes manipulation through training techniques: ballet, gymnastics, sports or work tasks. The goal is to manipulate our body in certain patterns for coordination, efficiency, aesthetic pleasure or competition. Throughout is our layered relationship to sexuality, usually the repression or redirection of sexual energy in conjunction with religious and cultural convention. There is confusion around all of the digestive functions, from eating to stomach growls to elimination, and a generalized "hush" about what is going on in the organs and the emotional centers of the body. Throughout our lives, but especially during adolescence, conformity to outer images of what the body is supposed to be, defined by social, cultural and religious norms, makes a division between our inner impulses and our outward manifestations. Less and less attention is given to what is coming from inside. We often need instruction on developing a healthy dialogue with our physical being. As young adults, much of the time is spent trying to "do" something to ourselves, to look better, get stronger, be thinner, work harder. And as mature adults and senior citizens, we are encouraged to deny or mask the aging process, to glorify youth rather than appreciate the beauty inherent at every age.

Consider the amount of time spent feeling good about our bodies. How often do we communicate with ourselves? Do we enjoy our physical capacities and efficiency? Our many years of schooling bring a separation of mind and body (sit still and learn). Cultural stereotypes and advertising emphasize the body as youthful sexual object. Physical training techniques and medical practices can lead to a view of the body as a machine, needing to be repaired by someone else when necessary. There is often a sense that one is either the master or the victim of one's own body. When communication breaks down, we are left polarized within ourselves. It becomes important to understand that the body has its own way of functioning, its own way of telling us what's going on inside, its own logic. Much of our task is to learn to listen. ❖

Constructive Rest
15 minutes

Lie on your back on the floor in a warm, private place.
○ Close your eyes.
○ Bend your knees and let them drop together to release your thigh muscles.
○ Let your feet rest on the floor, slightly wider than your knees (or prop pillows under your knees for support) and release your legs.
○ Rest your arms comfortably on the floor or across your chest.
○ Relax into gravity; allow yourself to be supported by the floor.
○ Feel your breath, and the responsiveness of the whole body.
○ Allow the organs to rest inside the skeleton (the lungs and heart, the digestive and reproductive organs); feel the contents released within the container.
○ Allow the brain to rest in the skull.
○ Allow the eyes to float in their sockets.
○ Allow the shoulders to melt towards the earth.
○ Allow the weight of the legs to drain into the hip sockets and feet.
○ Allow the surface of the back to move against the floor as you breathe; feel the ribs articulating at every breath.
○ Allow your jaw to gently fall open; feel the air move in and out through your lips and nose.

As you release your body weight into gravity, the discs are less compressed and the spine begins to elongate. You may need to lift your head or pelvis and lengthen the spine on the floor to accommodate this change. Constructive rest is an efficient position for body realignment. It releases tension and allows the skeleton and the organs to rest, supported by the ground. Constructive rest is useful at any time of day, but especially if done for five minutes before you sleep. The relaxation of the body parts returns the body to neutral alignment so that you don't sleep with the tensions of the day. Constructive rest is discussed by Mabel Todd in her book *The Thinking Body, A Study of the Balancing Forces of Dynamic Man.*

Note:

As we remember from playing blindfolded games in childhood, closing the eyes heightens the awareness of our other senses. As we relax our vision, we see in a different way.

Rolling up the spine

Transition from floor to standing
Three minutes

Lying in constructive rest:
○ Roll to one side of your body, allowing the head to stay relaxed on the floor. Feel the effects of gravity as you lie on your side.
○ Spread the palms of your hands on the floor and push into the floor to come to seated. Feel the change of gravitational pull as you sit in vertical.
○ Again, place both palms on the floor in front of you. Press into the hands and simultaneously rotate your pelvis off the floor so there is no weight placed on the knees. You are now in a relaxed push-up position with the pelvis in the air, knees bent, weight supported on hands and feet.
○ Relax your neck and walk your hands back to your feet, bending your knees as you need, so there is continuous flow.
○ Slowly roll up your spine, letting the weight drop down into your feet. Allow your head to hang forward until you reach the end of the roll up.
○ Feel the parts of your body balanced in relation to gravity.

This transition reduces stress on the knees and lower back. Repeat it a few times so the sequence is comfortable. Breathe naturally as you move.

Bodystory
Two hours
Give yourself time to collect as many memories as you can.

✎ Write a personal bodystory. Include:
• the story of your birth (pre-birth if possible; the health and activities of your mother affect life in the womb)
• your earliest movement memory (earliest kinesthetic sensation you can remember. Examples: being rocked, learning to swim, bouncing on your parent's knee, falling from a tree, riding a bicycle)
• training techniques (sports, dance, gymnastics, musical instruments)
• environment where you lived (mountains, plains, forests, oceans all affect how you move, how you perceive)
• comments you heard about yourself which shaped your body image ("Oh, what a cute chubby child! Stand Up Straight! He's going to be tall like his dad. Children are to be seen and not heard.")
• attitudes towards sensuality, sexuality; gender images
• injuries, illnesses, operations
• nutrition, relationship to body weight, strength, flexibility
• anything else that interests you.

PROPRIOCEPTION AND SENSORY AWARENESS

How do we register body position in space? Without looking at your body, take a moment to observe how you are sitting. How do you know where your feet and arms are in space, the tilt of your head, the curve of your spine? Throughout your body are sensory nerves with specialized receptors to record muscle stretch, pull on tendons, joint compression and the position of your head in relation to gravity. These nerves are referred to as **proprioceptors** ("self-receivers"), and they give us our kinesthetic sense. Proprioceptors are essential for movement coordination and thus maintain continuous input to the central nervous system for interpretation and response. Proprioceptive receptors can be found in the skeletal muscles, the tendons in and around joints, and the internal ear. **Muscle spindles** tell us about muscle length, **golgi tendon organs** detect muscle force and the pull on tendons, **joint receptors** monitor compression in our joints, and **maculae and cristae** in the inner ear apprise us of equilibrium. The receptors must transform a stimulus from the external environment into a nerve impulse to be conducted to a region of the spinal cord or brain in order for it to be translated into sensation.

The somatosensory cortex of the cerebrum has a precise map representing sensory information from all parts of the body, and works in conjunction with the cerebellum of the brainstem to maintain a continuous, cumulative picture of the body's position in space. The cerebellum, in particular, is responsible for constant coordination and correction of posture, movement and muscle tone. Even more fascinating, it holds the image of where you just were, where you are now, *and* it projects where you will go next. Remember the sensation of reaching for a stair with your foot when there was none? Your brain projection was different from actuality. (See the Nervous System, page 119) Stimuli carried to the spinal cord may initiate a spinal reflex – such as a knee jerk reflex – without input from the brain; those carried to the lower brainstem/cerebellum initiate more complex, subconscious motor reactions such as a reflexive postural shift to relieve muscle tension; sensory impulses which reach the thalmic level of the brain can be identified as specific sensations and can be located crudely on the body such as awareness of generalized pain or tension; those reaching the cerebral cortex can be located clearly on the body, such as awareness of position and movement, and connect with memories of previous sensory information so that the perception of sensation occurs on the basis of past experience.* For example, impact on the shoulder joint as an adult might provoke a protective response based on an injury from a fight in elementary school. Thus we carry a neuromuscular memory of our

Choice

A thirteen year old wildlife enthusiast was teaching me to handle a milk snake. I like snakes, but as soon as I saw its diamond-patterned body and flashing tongue, I tightened my muscles and stepped back. "The key to holding a snake is never to squeeze it or hold it too tightly," my young instructor informed me. I let my muscles relax and felt the diverse sensations happening throughout my body. Then I could see the snake more clearly and respond to its particular movements; I could act rather than react. In the moments between perception and response, I had choice.

❖

At a workshop with Nancy Stark Smith, one of the founders of Contact Improvisation, I was having trouble releasing my weight to be lifted by (or to lift) my partner. I held low level tension in my body all of the time to protect myself. "Tension masks sensation," she said to the class," and sensation is the language of the body."

Photograph: Erik Borg
Middlebury College Dancers

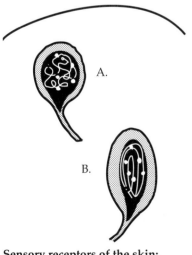

Sensory receptors of the skin:
A. Krause corpuscle registering cold
B. Pacinian corpuscle registering pressure and vibration

personal bodystory. Any body part can possibly be brought to our conscious attention with practice; it can also, fortunately, be left to provide information for body functioning without our awareness. Refined movement skills such as partnering in dance, an efficient tennis serve, and a surgeon's or pianist's hands in action, necessitate a highly developed kinesthetic sense.

The **general** and **primary** senses work in conjunction with the **proprioceptors** to monitor body awareness. The **general senses** include receptors for touch, pressure, vibration, cold, heat, and pain. They are located in the skin, the connective tissue and the ends of the gastro-intestinal tract; pain receptors are found in almost every tissue in the body. Visceroceptors, located in the blood vessels and organs, provide information about the internal workings of the body. Again, sensory information arises from the peripheral nervous system and is directed into the spinal cord, and then to higher centers in the central nervous system. If information reaches the highest level, the cerebral cortex, conscious sensation may occur. Some areas of the body such as the lips and hands are densely packed with sensory receptors, and others such as the trunk and thighs have few. Specific nerve ending receptors include: **Pacinian corpuscles** registering deep pressure and vibration; **Ruffini's end organs** for deep, continuous pressure and joint compression; **Merkel's discs, Meissner's corpuscles**, and **hair end organs** for light touch; **Krause corpuscles** for cold, **Ruffini corpuscles** for heat; **free nerve endings** for pain (and light touch).

The **primary senses** have specialized receptors for vision, hearing, smell and taste located in specific organs in the head (eyes, ears, nose and tongue). They project information to related lobes of the cerebral cortex: the occipital lobe, temporal lobes, and frontal lobe respectively. Awareness is selective: we can use our primary sense organs to listen for a baby's cry as we talk, or watch the expression on our listener's face, or smell bread baking in the kitchen, or taste the chewing gum in our mouth, or experience all of the above simultaneously. We choose where we focus our attention by our intention. As movements or stimuli become familiar, awareness of sensation diminishes. For example, I may feel a chair when I first sit down, but this awareness passes quickly. Nerve endings adapt, that is, they cease registering information or "firing," at different rates. Crucial receptors, such as those associated with pain, detecting chemicals in blood, or body position adapt slowly. The more developed and thorough our capacities for receiving and responding to sensory information, the more choices we have about movement coordinations and body functioning. ❖

* Gerard Tortora and Nicholas Anagnostakos, *Principles of Anatomy and Physiology*, pp. 344-345. For further information see Deane Juhan's "Skin as Sense Organ," "Touch as Food," and "Muscle as Sense Organ" in *Job's Body, A Handbook for Bodywork*.

Body scanning
20 minutes

Lying in constructive rest, or seated comfortably with your spine vertical: bring your awareness to the top of your head.

○ Observe, with eyes closed, any sensation you feel on the top of your head. It might be a tingling, a vibration, an itch, a pain. It might be a feeling of pressure, heat or cold, the touch of air on your skin.

○ Continue to observe any sensation you feel on the top of your head. (The repetition of the language helps to focus your attention.) If you feel nothing, just wait (while perception of your nerve endings gets more sensitive).

○ Bring awareness to your face and scalp. Observe any sensation without judgment; the task is to feel what is really happening in your body, without evaluating whether it is good or bad, pleasant or unpleasant. Experience your body just as it is at this moment in time.

○ Move your mind's eye to your neck. Remember to give equal attention to any sensation which you feel on your neck – tingling, the touch of cloth on the skin, your hair as it brushes the surface.

○ Continue to the right arm, the left arm, the back surface of the body, the front surface of the body, the pelvis, the right hip and thigh, the right lower leg and foot, the left hip and thigh, the left lower leg and foot. Bring your awareness to the soles of the feet.

○ Finish by observing your breath as it falls in and out of the nose and the mouth, moves the ribs, and stimulates the skin of the lower back and belly.

○ Slowly open your eyes; allow yourself to remain aware of sensation as you include vision.

Body painting
20 minutes

Lying comfortably on the floor, image a color of your choice which covers the surface you lie on.

○ Begin very slowly to paint your entire body with this color by moving your body surface in contact with the floor.

○ Be sure to touch every area of the body. Include: between the toes, scalp, eye sockets, behind the ears, under the chin, all surfaces of the pelvis, backs of the knees, wrists, armpits.

○ You may use another body part which is already painted to stroke hard to reach areas. Take as long as you need; do a final body scan to ensure every surface is covered in paint.

○ Begin a **proprioceptive warm-up** by following any impulses your body has for movement – a stretch of the arm or wriggling of the toes. Whatever feels good is "right."

○ Move nonrepetitively following the impulses of the muscles and joints to stretch and move.

○ Pause. Imagine yourself as a painted sculpture. Rest.

Note:

Each of us has areas of the body which we "shut off" for various personal reasons, including childhood trauma, embarrassment, injury, or neglect. These areas may be numb – hyposensitive – resulting in a lack of sensitivity, or the response may be heightened – hypersensitive – resulting in a "ticklish" sensation (nerve endings which detect no recognizable pattern from touch and are surprised, producing the familiar hysteria and giggling). Hyposensitive or hypersensitive areas can be brought into a balanced body picture through touch, proprioceptive warm-ups, and body scanning which relax muscle tissues, increase blood flow, and generally equalize attention to include all sensations.

Body scanning is also a component of Vipassana Meditation. For information about this form, write the Vipassana Meditation Center, Box 24, Shelburne Falls, Massachusetts 01370.

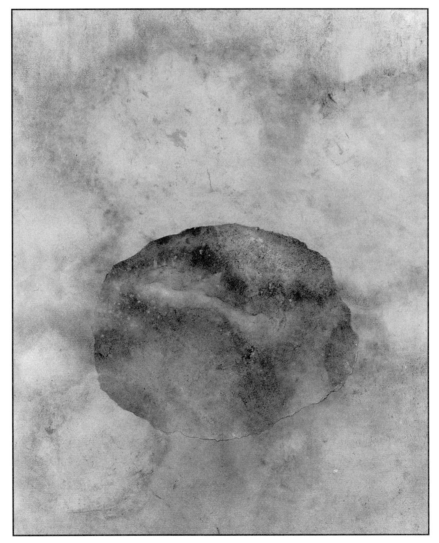

Collage: Rosalyn Driscoll
"Dawn Pilgrimage"

THE CELL

The cell is the fundamental unit of the body. The abilities of the cell to reproduce, to metabolize, and to respond to its environment, are basic to human life: creativity, processing, and responsiveness to change.

Cells have common properties but vary according to function in the body. Each cell is composed largely of water, the basic substance of the body. Water is contained within the **cytoplasm** of the cell which is concerned with metabolic activities such as the use of foodstuffs and respiration. The **cell membrane** differentiates the cytoplasm from the surrounding external environment and creates a semipermeable boundary governing exchange of nutrients and waste materials, and responding to stimulation. The **nucleus** supervises cell activity. The forty-six chromosomes in each human cell contain the genetic code for the individual body and for the specific functioning of each cell. Each nucleus, therefore, contains a master plan of the whole body. Individual cells have different functions; for example, a muscle cell contracts, a nerve cell transfers electrochemical signals, a fiber-producing cell produces connective tissue fibers. A collection of like cells of similar structure and function is called a tissue. Groups of coordinated tissues form structures (organs), which comprise a body system. For example, bone cells form bone tissue, which makes bones, which create the skeletal system.

The cell is the functional unit common to all body systems. For our study, we will differentiate seven body systems as defined by Bonnie Bainbridge Cohen: skeletal, muscular, nervous, endocrine, organ, fluid, and connective tissue. Although we can look at each system individually, it is important to remember that the body functions as an interrelated whole and that the systems balance and support each other.

The human body develops from the union of two cells, the sperm and the ovum. The fertilized ovum divides repeatedly to create many cells. Within a day or so of fertilization, the ovum differentiates into embryonic tissue layers: the ectoderm, endoderm, and mesoderm in that order. The ectoderm is the origin of all nervous system components and the skin; the endoderm, of the digestive tract and organs; the mesoderm, of the connective tissues (blood, bone, muscle, ligament, tendon, fascia, and cartilage). Thus, the skin begins from the same embryonic tissue layer as nerve tissue; the blood originates with connective tissue. Cells vary in their adaptability to change or healing, and in their rate of reproduction. For example, some skin cells reproduce through cell division daily, while a nerve cell may remain for a lifetime and heals slowly if at all. The skin is the external membrane of the body, a highly sensitive boundary between our body and our environment. Sixty to seventy per cent of lean body

Exchange

While attending workshops about experiential anatomy, I would sometimes get stuck. I put a barrier to learning around absorbing new information because I felt threatened by change and by opening to a group. One particularly tense day, the teacher talked about the cell. She described osmosis and the movement of fluids through the cell membrane. She looked around the room and said, "Remember, change is only a membrane away." The tightness of my skin dissolved. Once relaxed, I could allow information to come and go, keep what was useful, and express my own ideas. By releasing my outer membrane, I could allow exchange.

❖

An instructor once said to a group of adults, "See if you can walk through the room without feeling responsible for anyone." And also, "Within the cell, feel the movement in stillness. Within the group, feel the space in closeness."

weight is water, and the skin literally keeps us from drying up. Two-thirds of this water is within the cells (intracellular) and one-third is between the cells (extracellular). The skin also maintains body temperature (through sweat), contains receptors of various sorts, and provides a responsive, protective covering. It forms orifices such as the mouth, the nose and the anus, leading to the passageways of the digestive system and the respiratory system which can be seen as extensions of the external environment.

Oxygen, essential to cellular life, comes into the body through our nose and mouth and travels through the trachea to the lungs. As the diaphragm descends, the lungs are expanded by the inrush of air called inspiration. When the diaphragm releases, the lungs are compressed to expel carbon dioxide in a process called expiration. The oxygen is absorbed through the capillaries in the lungs and enters the blood to be pumped by the heart throughout the body. Arteries carry the oxygen-rich blood from the heart to the periphery. Each cell participates in the absorption of oxygen and the removal of waste materials in connection with a process called cellular respiration. Deoxygenated blood returns via the veins to the heart. Through this process, every cell is in connection with the outer environment and "breathes."

Tension in any part of the body restricts cellular activity vital to healthy tissue. Through bodywork, we use the responsiveness of the cell membranes and the skin to heat, vibration, and touch to bring awareness and affect change. ❖

Cellular breathing
20 minutes

Lying in constructive rest with your hands on your ribs, eyes closed:
❍ Bring your awareness to your breathing. Feel the air coming in through your nose and mouth, passing down through the trachea in your neck, and filling the lungs inside your ribs. Feel all the ribs move as you breathe. In the damp, warm environment of the lungs, the oxygen is transferred from the air to the blood through tiny capillaries. This is " lung breathing." Three-fifths of the volume of your lungs is blood and blood vessels.
❍ Feel the pulsing of your heart. Image the blood being pumped by the heart through the arteries, carrying oxygen from the lungs to every cell in the body. This absorption of oxygen and removal of waste materials through the cell membrane is called "cellular breathing." Image the deoxygenated blood returning through the veins to the heart, and the process repeating. Place your hands on your belly and your ribs, and feel them both move as you breathe.
❍ Image the flow of oxygenated blood from your heart down into your belly. Let this continue through the hips and knees, and into the ankles and feet. Allow the flow to return like a wave from your feet to your heart. Feel the movement under your hands. Image the flow of fluids moving from your heart up through the neck and into the skull to bathe the brain, and back to the heart. Feel the flow out through your shoulders and elbows and hands, pooling in your palms and fingers and returning to center. Feel the continuity and constancy of flow through your whole body. Image the fluids moving simultaneously from center to periphery and from periphery to center.

Breathing spot
5 minutes

In constructive rest: roll to your side, flex arms and legs close to the body and continue to roll to a "deep fold" position: arms and legs tucked close to body, forehead on floor, spine curved.
❍ Place your hands on your lower back, just above your pelvis. Feel the movement of the skin and muscles as your breath enters the lungs and is released. The diaphragm compresses the abdominal organs and expands the back. We can call this area your "breathing spot." Encourage its movement with each breath.

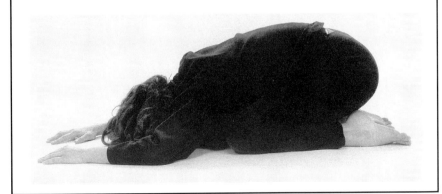

Cellular awareness:

Image yourself as a single cell. Feel the boundary of the outer membrane. Be aware of yourself contained, as a single unit, with all parts of the body contributing to the whole. Allow exchange with the world around you. Notice what flows in, what flows out.

Photograph: Bill Arnold
"Allan's Boys"

EVOLUTIONARY STORY:
In the Water

The evolutionary story of the human species begins in the water. In that fluid environment, diverse forms of life emerged. Plant life preceded animal life as a producer of oxygen and a primary food source. About 3.5 billion years ago, according to available fossil records, photosynthetic bacteria and blue-green algae formed in the primordial seas demonstrating the basic characteristics of life: the ability to reproduce, metabolize, and respond to change. By 1.5 billion years ago, the first single-celled organisms appeared with the capacity to reproduce sexually (mixing of the DNA of two cells) instead of by cell division (division of one cell into two identical parts) thus increasing the potential for diversification. Around 600 million years ago, we find evidence of a full range of multicellular life with unique body structures: primitive forms such as sponges and jellyfish; shelled mollusks including snails and clams; starfish and sea cucumbers with radial (five fold) symmetry; elongated or segmented forms of worms; joint-limbed creatures whose contemporary ocean relatives include lobsters, shrimp and crabs; and the more advanced creatures with backbones such as cartilaginous sharks or bony fishes. We can experience a similar diversity today by observing the multiplicity of underwater life present in a coral reef!

Throughout the evolutionary story, species (populations of individuals which habitually mate with one another) become extinct, remain constant, or adapt in a process called natural selection. Change occurs by random genetic mutation, followed by successful reproduction. Some creatures evolve characteristics which allow survival when new conditions such as increased population or climatic shifts force a change. One example is the freshwater lungfish, which evolved internal air bladders and muscular lobes on its fins. These preadaptive structures eventually facilitated the species in walking on land. Some forms, such as the shark, are so successful that they have remained almost the same for millions of years. Others, like the dinosaur, dominated the land for 150 million years before they became extinct. The human species, *Homo sapiens*, has walked the earth for a mere two million years and continues to evolve.

We will look at three types of body symmetry and their relationship to human movement patterns. **Asymmetry** is characterized by a single-celled organism, like an amoeba, in which all components of the membrane are of equal importance. Exchange of nutrients and waste materials takes place through this semipermeable membrane, and the organism moves as its protoplasm shifts in relation to the environment. A sponge, a loosely organized collection of single cells, provides a multicelled example of asymmetry. (If you put a sponge through a sieve, the cells will reassemble in their original form.) **Radial symmetry** brings the mouth and gut to the

Layers

I was backpacking into the Grand Canyon from the north rim. Our descent into the deepest cleft in the earth's surface took three days. In our group was a geologist who loved reading history through the layers of stone. We began by looking for cockleshells and other fossils in the white bed of Kaibob limestone crumbling under our feet – an estimated 225 million years in the making. On the first night, we camped facing the smooth Redwall – limestone 300 million years old; on the second night we slept at the site of an old Anasazi ruin, facing what was called the Great Unconformity – an eroded era of ancient rock, formed before life was present on the earth, now missing from the geological "book" of the canyon walls. On the third day, as we lay in the hot sun on the banks of the Colorado River, my friend pointed to the black metamorphic rock which lined the edge of the water, and said, "That's Vishnu shist. Two billion years." Then he added, "Now that's old."

❖

During a conversation with a college biology professor, I said,"I encourage students to study the human body." He responded, "I remind them that humans aren't the only species on earth."

I was traveling in East Africa. We were in Amboselli National Park in Kenya, on a game drive at dusk. We had stopped to watch elephants walking parallel to us in the trees. As if on cue, the elephants turned and made a path directly towards our van. They formed a straight line, with the mother and father on either end protecting two babies of different ages between them. As they progressed systematically, we backed up our van, and they passed a few feet in front of us without ever breaking stride. Their eyes looked ancient, the skin sagged off huge bodies, and the trunks uprooted turf and searched for minerals in the soil and rock. They had been around for a long time. I imagined them saying, "Move over."

We were in Tanzania, gazing from a dusty overlook towards Olduvai Gorge. We were amidst a small gathering of tourists from around the world all wearing safari hats and sipping sodas. It was here that noted Kenyan paleoanthropologists Mary and L.S.B. Leakey discovered fossil hominids – an Australopithecine skull, "Zinjanthropus" in 1959, and fragments from Homo habilis in 1961 – dated 1-2 million years old. Nearby, Mary found footprints preserved in volcanic ash determining that as early as 3.7 million years ago, upright, bipedal forms of the species Homo erectus lived in Eastern Africa. We listened to our guide translate this earliest history of the human heritage into various languages. There was a sense of Africa as common origin.

center, with appendages radiating from this core. The starfish, as a lasting example, walks on the ocean floor to find and devour its food. **Bilateral symmetry** differentiates a "head" and a "tail" end with paired body parts. The mouth and primary sense organs gravitate towards the head, and appendages for propulsion and elimination towards the tail. In some species, like the shark, a cartilaginous "spine" links head and tail for stability and directionality, with fins used for buoyancy (if a shark stops swimming, it sinks), but in others such as the bony fishes, a skeleton and paired appendages develop for additional stability and mobility in the water (with internal air bladders providing buoyancy). The sea squirt demonstrates the transition from invertebrate (without a spine) to vertebrate form. The adult has radial symmetry and lives planted in the ocean floor, but the juvenile form has a notocord (primitive spine) and swims like a tiny fish, with bilateral symmetry, before changing into its less mobile adult form. Bilateral symmetry and a bony skeleton provide efficiency for basic survival needs such as going towards food, escaping from enemies, pursuing a mate, and exploring the environment, and are useful components in the transition to land. Although our discussion is primarily anthrocentric – focused on the human species – ninety-five per cent of the animals on earth today have no backbone!

Just as we carry the saline solution of the ocean in our blood, our structure holds the possibilities of earlier forms of body symmetry. We can move asymmetrically, as in our early morning yawning and stretching, allowing our skin and proprioceptors to be our primary sense organs before the cerebral cortex (the newest portion of the brain) directs our awareness. Contact Improvisation and Authentic Movement are techniques which focus on stimulating and responding to all surfaces and structures of the body equally. We can move in radial symmetry, like the starfish, or the Leonardo da Vinci drawing of "geometric man" with body parts radiating from the solar plexus. Cartwheels demonstrate this symmetry. Martial arts also organize movement around the "belly brain" (autonomic nervous system) focusing on the tant'ien (in tai chi) or the hara (in karate) for "centered" energy. We can move in bilateral symmetry, undulating our segmented spine like fishes or whales. Kundalini yoga and many primitive dance forms use this powerful source of head to tail integration. Although our outer form is organized bilaterally, many of our internal organs (such as the heart, liver and intestines) retain their asymmetrical form. The skin, our largest organ, remains a primary receptor for receiving stimuli from the environment. Our vertical posture dominates our contemporary lives, but the structure of our bodies lies in our evolutionary past.

As we tell the evolutionary story, we are reminded that there is more to the universe than the human mind can grasp, and that we seek models to try to understand our existence – scientific, religious, and artistic models among others. We can relate what we know from a scientific perspective (primarily from fossil record) about the origin of life and the development of the species, but fact goes hand in hand with mystery. When a new discovery is made, the whole picture shifts. Thus, our information is constantly changing, and the insights we have as individuals are part of the discovery. ❖

Body symmetries
20 minutes

Lying on the floor on your belly, arms and legs spread in an X: Slowly lift one arm about an inch off the floor, feel its weight, and let it return to the floor. Repeat with a leg, your head, then the pelvis. Let all of your weight rest into the floor.

○ **Asymmetry:** Imagine that you are suspended in water. Begin moving slowly, brushing all surfaces of your body against the floor or another body part as though stimulated by the water. You are free from gravity. Continue moving nondirectionally, following impulses for movement and responding to sensations. This form of body symmetry is called asymmetry. An example would be a single-celled amoeba suspended in the water today. In this symmetry, no one part of the body surface has priority over another part.

○ **Radial symmetry:** Shift your awareness to your belly. Imagine that your mouth and your brain are in your belly, and your head, arms and legs are appendages radiating from this center, like a starfish. Extend your tail as well (the end of your spine). Stretch yourself out in the X, and imagine or feel the connection through the center of the body between your right hand and your left foot, your left hand and your right foot. Your head and your tail. (If you are with a partner, have one person lie in the X, and the other gently press each appendage along its axis towards center like the spokes of a wheel. The compression and release of the bone ends moves the belly, the hub of the wheel, as you lever from periphery to center. Do the head and the pelvis as well as the arms and legs.) Continue lying in the X on your belly or your back, and reach one arm diagonally across the body, connecting into your center and rolling you to your other side. Move slowly so you can feel the rotation move through your spine. Repeat with a leg: reach a foot diagonally across the body until you feel the pull into center, and let the movement roll you to your opposite surface. Repeat with the remaining limbs.

○ **Bilateral symmetry:** Begin to focus your mouth and primary sense organs towards the head end of the body. Follow your mouth and feel its connection through your body to your tail. Follow each sense: Smell, taste, sound, sight and follow them into locomotion. ("Hear" and "smell" with your whole body as well, like a fish. "See" light all over you, like a worm.) Go towards a smell, move away from a sound. Imagine propelling yourself through the water for food or to avoid an enemy. The long spine connects the sensory end of the body, the head, with the motor end, the tail. The "arms" and "legs" are fins used for stabilization more than locomotion. Imagine that you have been washed up on shore. Gravity, friction, inertia are now part of your environment. The sun is hot, predators may be near. What forms of locomotion might you devise to move to safety or towards food, shelter, or a mate in this new environment? Explore the many choices, keeping your senses alert for survival. Finish your work. You are now ready for the evolution to standing!

Killer whale, from a decorative hanging of American Eskimo art

Painting: Gordon Thorne
"Study for Woman"

EVOLUTION TO STANDING: On Land

As our ocean forebearers found themselves on the shore, a new relationship to gravity began. Inertia and friction of body parts against the earth, and the need to lift the head up for breathing and perception required changes of body structure. Fins evolved into weight bearing appendages. In amphibian and reptilian species, they gradually folded in towards the center of the body to move the belly off the earth or disappeared altogether, as forms of **crawling** developed. We see various approaches in the successful adaptations of salamanders, lizards and snakes. The salamander maintains belly-down locomotion, lizards pump their torsos off the ground in the familiar "push-up" motion, and the snake has no appendages. If we observe a lizard carefully, we can see the distinction between the rougher "back" surface exposed to the elements, and the lighter "belly" surface in contact with the ground. As the body parts rotated further towards center, the "back" of the legs became the front. The "elbow" of the foreleg faced backwards to prop up the body; the "knee" joint faced forward for propulsion and support; and the "hand" rotated palm-down for additional leverage against the ground. In the human, the hairy surface of our thighs was once the "back" of the legs and the bones of our forearms cross to allow mobility at the wrist.

The evolutionary tree continued to branch into diverse species, each with particular characteristics for survival. Following evolutionary theories, our mammalian forebearers survived as quadrupedal, ground-dwelling insectivores and herbivores as larger and carnivorous animals made life threatening. Some species moved to the trees and adapted to a posture of **hunkering** on branches, squatting with feet grasping the branch and legs folded (flexed) to the belly, leaving the hands or forepaws free for eating, grooming and gesturing. The new use of the hands developed a crosspattern of thumb opposite fingers for grasping, and coordination of eyes with hands. Detailed movements of the fingers, toes, and mouth (we have more muscles in our face than any other animal) refined the nervous system and prepared the way for communication through gesture and language. The hunkering posture, our familiar "squat," encouraged three structural changes in the body: the heel bone (calcaneus) gravitated to the back surface opposite the toes, an arch was formed by the connecting bones of the foot, and the pelvis shortened to allow the femurs to fold in squatting position. Our ape ancestors began hanging from their arms and swinging from limb to limb through the treetops, called **brachiation**. This allowed extension of hip and shoulder joints, repositioning of the scapulae to the back surface of the body for hanging and lateral reaching, contralateral rotation at the waist necessary

Language

I was in China, performing and studying. It was my first experience in a country with a 3000-year cultural history. I had bought a snake drum, which I carried slung over my shoulder, and was traveling to various areas. In beautiful Guilin, I took a boat trip down the Li-jiang river. When we landed, the tourists on the cruise were besieged by local residents selling antiques and replicas, hats and bananas. Hands grabbed us, using body-pleading techniques of street salesmanship since no common language was spoken. My fellow tourists raced for the waiting buses, and I, like the others, became intent on escape. As I passed through the faces, I saw a man under a tree selling a musical instrument. I gestured to him to play some music, but he pretended he didn't know how. I began playing my drum. Everyone stopped what they were doing. I began to dance. I watched the face of an old woman staring at me, toothless but elegant. Suddenly she started smiling and clapping, encouraging me to go on. Soon both travelers and residents were mingling and laughing. As we left, the man under the tree was stringing his instrument. This was my first clear experience of sound and dance as language.

27

Body as Home

A good friend, whose parents have both passed away, was speaking of the grieving process. She said that she felt a sense of dissolution of home, even though she was married, owned a house, and had a community of friends. Do you think of your body as your home? I asked.

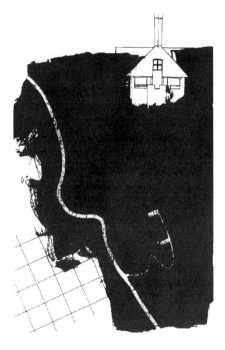

**Architectural drawing:
Whitney Sander**

for swinging, and elongation of the spinal curves. The lumbar curve (our "lower back") formed last, after the arboreal "swinger" returned to the ground as a semiquadriped. The transition then to upright posture was accompanied by the anterior (forward) curve of the lumbar segment. These characteristics prepared the way for two-footed, vertical posture: a bipedal stance.

Two-footed alignment involved a high center of gravity over a small base of support (the feet). The vertical axis was constantly swaying over this base, requiring contraction and release of the muscles of the lower legs to keep on balance. A subtle shift past the base initiated walking, striding, or running. In effect, they were constantly falling. Instability is basic to our bipedal stance.

Vertical alignment brought new possibilities for life on the ground. The preparation of the foot, with heel opposite the toes, and arched tarsals and metatarsals, gave propulsion and shock absorption for a **striding gait**. (See the Arthrometric Model, p. 114) Rotational abilities at the waist increased the capacity for contralateral swing of the arms opposite the legs, offering heightened balance and agility. Without moving their feet, they could reach in front, to the sides, or behind. This three-dimensional rotation of the spine and the ball and socket joints permitted quick response in any direction, our multidimensional agility. With the addition of meat eating and hunting, the body adapted to the need for attack and retreat. The free-swinging pelvis held the organs in a lightweight bowl and allowed mobility (unlike our close relative the gorilla, who evolved from a common ancestor but whose large pelvis restricts vertical extension and whose knuckle walking and vegetarian diet encourage a more passive existence). The use of tools and the development of family groups increased the need for articulation and communication. In essence, they could stand, move three-dimensionally, manipulate their environment and articulate with gesture and sound. Neurological adaptability increased the potential for survival.

The basic characteristic of our species, ***Homo sapiens*** (wise man), is the increased capacity of the brain. As all of our systems developed simultaneously with our skeletal-muscular changes, our physiological capabilities were matched by our capacity for three-dimensional thought in the past, present and future. We are able to reflect on where we have been, contemplate where we are, and plan where we are going. This capacity for reflection, planning, and manipulation of our environment brings the responsibility of choice. Our ability to plan and to shape our environment makes us responsible for what we create and for how we choose to live in that creation. Thus, our responsibility is to remain able to respond, moment by moment, to the choices which occur. ❖

Evolution to standing
30 minutes

○ Lying on the floor in constructive rest. Review asymmetry, radial symmetry, and bilateral symmetry. 10 minutes.

○ On the shore, feel your new relationship to gravity. Lying on your belly, keep the elbows and knees "locked" and attempt to lift your body from the shoulders and hips.

○ Gradually draw the appendages (once your fins) towards the center line of the body, rotating fingers and toes forward, palms down, similar to a salamander or lizard. Explore crawling.

○ Find ways to lift the belly off the earth to reduce friction. You may eventually be walking in a four-footed posture like a squirrel. Imagine yourself to be a small, four-footed mammal, heading to the trees for protection from larger animals.

○ Begin to sit, hunkering on branches. The pelvis hangs between the legs in a familiar squat position, the feet wrap around the branch for support, the hands are free to groom yourself or others, to feed, and also to locomote through the tree tops. Begin hanging from a tree branch by your arms, like a monkey. Imagine that the floor has disappeared. The legs are elongated from their flexed position in hunkering and are dangling free. The shoulder joints are stretched and the spine lengthened. Let yourself swing from your arms reaching one after the other to propel you through space, brachiating like when you were a child swinging from bar to bar on jungle gyms. As you speed up your movements, you will feel a natural progression from homolateral movement (right hand swinging with the right leg, meaning same-sided) to the more powerful contralateral swing (the left leg swings across as the right arm reaches, opposite-sided). Feel how this movement pattern, brachiation, develops the rotation at the waist. Alternate brachiation with hunkering.

○ When things become peaceful, climb back down to the earth. We now have many options for movement. We can continue to hunker and squat, we can return to four-footed movement, or we can use the elongation of our hip sockets and the contralateral swing of our arms and legs to stand up and walk. To make the transition, motivate your movement by the desire to travel through space. To rise, feel the alternating, oppositional pulls of head and tail for movement towards vertical. Then begin the walk by reaching with the hand, as we did in brachiation. Because of our hunkering posture our feet are prepared with an arch and a toe opposite heel leverage for long distance walking, or the "striding gait" so unique to humans.

○ Walk around the room, feeling the heel of the foot reach as the leg swings to the front, and the toes (especially the big toe) push off vigorously in back. Let the arms swing, and use all the senses to inform your choices for survival. Let your walk develop into a run, and feel the spring of the arch cushioning your gait. Pause.

○ Stand quietly still. Close your eyes and feel the constant postural sway, forward over the toes and back to the heels, that keeps us balanced over our base of support, our feet. Open your eyes slowly and see the room around you from your vertical stance.

Rock painting in Kakadu National Park, Australia

29

Photograph: Bill Arnold
"Eiffel Tower"

BODYMEASURING:
Terminology

We live in a multidimensional structure. Our height, depth, and width help to describe volume in space. It is easy to think of ourselves as flat; mirrors and photographs give us the illusion that we are two-dimensional. Instead, we have sculpted fullness, and the curves and angles give force and agility to our body.

Planes, axes, and the **center of gravity** provide a common language for body measurement and movement description. **Planes** identify dimension in the body and divide it front to back, side to side, and top to bottom. (See illustration, next page) A primary plane divides the body equally by weight. A secondary plane is any plane parallel to the primary plane. The primary **transverse** plane divides the body equally by weight top to bottom. This plane passes horizontally through the center of the body, like a table top with the bust above and the pelvis and legs below. (A secondary transverse plane might pass horizontally through the knee.) The primary **sagittal** plane intersects the body equally between the right side and left side by weight. (A secondary sagittal plane might pass through the leg and divide the right side of the leg from the left side of the leg.) The primary **coronal** (or frontal) plane divides the body equally by weight front to back. (A secondary coronal plane might pass through the nose and toes.)

An **axis** is a line derived from the intersection of any two body planes. If we identify the line formed from the intersection of the primary sagittal (dividing right and left) and the primary transverse (dividing top and bottom) planes, we see that this line passes from the front to the back of the body, measuring depth. It is called the **anterior-posterior axis,** or the a.p. axis. An a.p. axis at the knee would measure the depth of the joint from front to back; at the head, it would measure the depth of the skull. Sometimes it helps to use sheets of paper to visualize three-dimensionally: place two sheets of paper beside the body as the sagittal and transverse planes. They will form a 90 degree angle where they would intersect. The line created where the two papers meet represents the a.p. axis and measures depth. Using your papers, identify the line formed from the intersection of the primary transverse (top and bottom) with the primary coronal (front and back). This line measures the width of the body and is called the **horizontal axis.** Find a secondary axis at your skull. Identify the line formed from the intersection of the primary sagittal and primary coronal planes. This axis measures the height of the body and is called the **vertical axis.** Find this with the two papers representing the planes. Find a secondary vertical axis in your leg.

Depth

Many college-age students are involved in getting ahead. Literally and figuratively, their bodies strain forward and their front surface, how they appear, is of prime importance. In their senior year, students take the experiential anatomy course. Together we work to find their back surface and to explore their depth.

❖

For many years I thought of myself as tall. At 5'3," this is not the case, but my closest friends were tall and I imagined myself at their height. This was useful as a dancer, and often after a concert a surprised audience member would say, "You look so much taller on stage!" During this same time, I was fighting with my body; I was always working at dancing. One day I thought of myself as short. I lifted my leg from my hip with ease, I stretched my arm from my shoulder without strain. Movements that had always been hard were accomplished with relative ease. By accepting my structure, I was free to move.

❖

When I was taken to the communal baths in Japan, I would see bodies of every age, from childhood to the edge of death, being touched by water. One old woman in particular would stand for a long time with the fountain pouring down her front. Next to her, my friend Mayumi was scrubbing my back. Inside myself, I felt the space between front and back.

The plane not involved in the formation of an axis is called the **plane of motion**. The body or body part **rotates** around its axis through the plane of motion. A movement, for example around our vertical axis (measuring height) would be a spin, or a head shake no. A movement around our horizontal axes (measuring width) would be a forward bend from the waist, a flip-flop, or a head shake yes; a movement around our anterior-posterior axes (measuring depth) would be a side bend from the waist, or a cartwheel. A movement which crosses the body planes, such as a spiral or a tennis serve, is considered an integrative movement, as all three planes are utilized. In the human species, our capacity for rotation at the waist, through the horizontal plane, gives us the potential for multidimensional agility. We can reach in any direction in space without changing our base, by spiraling around our center. There are many such spirals inherent in the body structure. Each muscle fiber, the strands of DNA, and the heart are in spiraled configurations; the fascia wraps and interweaves throughout our body. Our capacity for multidimensional, integrative activity is inherent in our structure.

The intersection of all three primary body planes creates a **point**. This point is the **center of gravity** around which all movement is organized for balance and mobility. In the human body, as we have noted, the center of gravity is generally between the fourth and fifth lumbar vertebrae on the front of the spine (behind your belly button). Move your center of gravity slightly; feel the body respond to keep balance. Move the center of gravity further in space. Feel it pull you into locomotion. Falling, walking, running, hopping, skipping, jumping, leaping are all locomotor patterns common to our two-footed structure.

Basic locomotor patterns move through the sagittal plane (with some rotation at the waist through the transverse plane to accommodate balance). These "sagittal movements" reflect our bilateral symmetry, with paired body parts and forward/backward orientation, like a horse. Cartwheels and side leans are "coronal movements" because they pass through the coronal plane. They reflect our radial symmetry. Spinning and turning are "transverse movements" because they occur through the transverse plane around our vertical axis (even when done on the ground as rolling). Spiraling movements pass through all three planes and integrate the dimensions of the body. ❖

Planes, axes and center of gravity

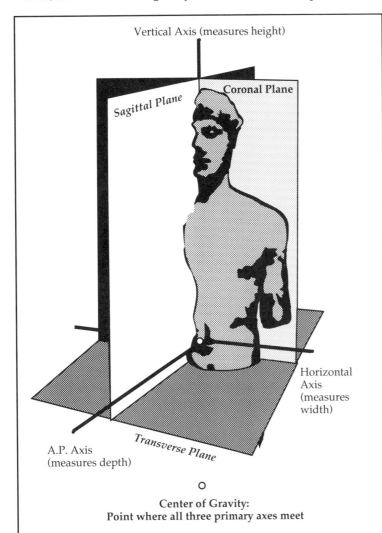

Vertical Axis (measures height)

Coronal Plane

Sagittal Plane

Horizontal Axis (measures width)

A.P. Axis (measures depth)

Transverse Plane

o

Center of Gravity:
Point where all three primary axes meet

Measuring the body
10 minutes

Lying in constructive rest: Extend your legs long on the floor. Bring your awareness to your forehead. In your mind's eye, drop a plumb line from your forehead to the floor. Feel the depth of your skull. Bring your awareness to your ribs. Drop a plumb from the front of your ribs, the sternum, to the floor. Feel the depth of your ribs.
○ Bring your awareness to your belly button. Drop a plumb from the front of the belly to the floor. Feel the depth of your pelvis.
○ Bring your awareness to the front of your right knee. Drop a plumb from the front of your knee to the floor. Feel the depth of your knee joint.
○ Bring your awareness to the toes of your right foot. Feel the length of your foot from the toes to the heel. The anterior-posterior axis measures depth of the body.
○ Return to the head.
In your mind's eye, draw a line from your right ear to your left ear. Feel the width of your skull. Draw a line from your left shoulder to your right shoulder.
Feel the width of your shoulder girdle. Draw a line from your right greater trochanter to your left greater trochanter.
Feel the width of your pelvic girdle.
Draw a line from the outer surface of your right knee to the inner surface of your right knee. Feel the width of your knee joint.
Draw a line from the outer malleolus of your right foot, to the inner malleolus of your right foot. Feel the width of your ankle. The horizontal* axis measures width of the body.
○ Return to the head. Draw the plumb line from the top of the skull to the base of your feet. The vertical* axis measures height, length of the body. Feel the three-dimensional movement of the ribs in breathing.
○ Repeat standing.

* Terminology is relative to standing anatomical position: body erect, arms at sides, palms forward.

Planes, axes, and planes of motion

Standing: Do a movement around your primary vertical axis, such as a headshake "no." The axis is formed from the intersection of the sagittal and coronal planes; the body rotates around the axis and moves through the transverse plane.
○ Do a movement around your primary horizontal axis, such as a bend from the waist or a summersault. The axis is formed from the intersection of the primary coronal and horizontal planes and the movement occurs through the sagittal plane.
○ Find a movement at a secondary horizontal axis; this might be a headshake "yes," a bend of the knee, a flexion of the ankle.
○ Do a movement around your primary a.p. axis. This might be a side bend, a cartwheel. An a.p. axis is the line formed from the intersection of sagittal and transverse planes. As you do your cartwheel, you are moving through the coronal plane. A side leg lift would be an example of movement around a secondary a.p. axis through the coronal plane.
○ Find a movement which passes through all the planes.

Other Terminology:

Abduction: Movement away from the body in the coronal plane. *Lift your arm to the side.*

Adduction: Movement toward the body in the coronal plane. *Lower the arm from the side.*

Flexion: Bending or closing movement of a joint or spine. *Bend the elbow.*

Extension: The act of straightening or opening out a limb. *Straighten the elbow.*

Proximal: The portion of an organ or structure nearest its base point or the trunk. *The shoulder is proximal to the elbow.*

Distal: The portion of an organ or structure furthest away from the base point or your trunk. *The knee is distal to the hip.*

Rotation: Movement of a bone around its own axis. *Turn your head to the right.*

Circumduction: Movement at a joint in which the distal end of a bone moves in a circle while the proximal end remains fixed. *Circle your arm.*

Medial: Nearer to the midline of the body. *Your big toe is on the medial side of your foot.*

Lateral: Further from the midline of the body. *Your little toe is on the lateral side of your foot.*

Homolateral (Ipsilateral): Related to same side of the body. *Walk reaching with your right arm and your right leg.*

Contralateral: Related to opposite sides of the body. *Walk reaching with your left arm and right leg.*

Sculpture: Michael Singer
"Sangam Ritual Series"

THREE BODY WEIGHTS AND POSTURAL ALIGNMENT

Postural alignment involves the three primary body weights, the skull, thorax, and pelvis, organized around a vertical plumb line. Our high center of gravity is delicately balanced over our base of support, the feet. If we draw three ovals, representing the three body weights, and connect these with a vertical axis from the top of the skull to the feet, we have a diagram of postural balance in the body. On our drawing, add a nose and toes to identify the front of the body, and heels for the back. The front is also called the anterior or ventral surface, the back the posterior or dorsal surface. We use labels of body facings to clarify communication. To further identify terminology, we are looking at the body from a lateral, or side, view. Anything in the direction of the head is said to be at the cranial end of the body, and anything in the direction of the feet, is caudal (of or related to the tail) from the center of gravity. A picture of the body shown from above would be a cranial view, one from below would be a caudal view. To locate the center of gravity, we identify the point created by the intersection of all three primary body planes, each dividing the body equally by weight (top to bottom, front to back, side to side). The center of gravity lies behind the belly button at the front of the spine (4th or 5th lumbar vertebra in most bodies), and intersects the plumb line.

To identify your own postural alignment, remember that the vertical plumb drops from the top of the skull to your feet through the three body weights. Because our body connects to the earth at the feet, we generally begin work on alignment from the ground up. We balance each body part in sequence. On the exterior of the body, the landmarks for alignment are: the side of the ankle (the malleolus), with the side of the knee, lined up with the side of the hip (greater trochanter of the femur), with the side of the rib cage and shoulder girdle (glenoid fossa), and the center of the ear. Tapping the head on the top of the skull helps to stimulate awareness of the vertical pole. ❖

Balance

Often when I am teaching a dance class, I will see a student in perfect balance. It is a beautiful thing – calm, simple, direct. And then they will recognize the moment and move slightly off center. Balance through the bones produces minimal feedback in the nervous system. It is a curious state of no-sensation. They move to feel themselves struggling, working, trying. And then they fall and look at me with frustration and say, "I can't balance."

❖

Two six-year-olds dance with me each week. At first, balance on their toes was part of their natural vocabulary. After a few weeks they began to mock falling down, making exaggerated faces, waving their arms wildly.

❖

In relationships, when something is quite balanced – nothing is happening – I get nervous.

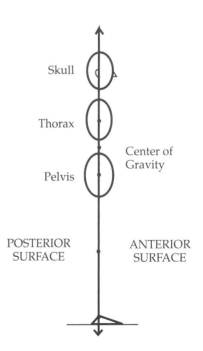

Three body weights: lateral view   35

Cone head and tail
10 minutes

Function: To feel postural alignment with dynamic pulls towards gravity and towards verticality.

Standing in plumb line, feeling your vertical axis: Tap the top of your head and imagine a weighted string dropping down through the center of your body from this spot until it touches the earth between your feet. Relax your weight through the bones into gravity; image a great tail extending down your spine into the center of the earth.

○ Bring your awareness back to the top of your head. Image centrifugal force spinning your body out into space away from gravity, while your feet remain reaching to the earth.

○ Image your plumb line growing above your head one, two, three inches until it extends, inch by inch, thirteen inches above your skull. Imagine that you have a cone head, like a tall pointed hat, that extends to this point. Inscribe a tiny circle on the ceiling from the top of your cone. Reverse the direction. Take a walk leading with your cone head. If working in a group, nod at others from the tip of your head. Come to standing. Feel the plumb line balanced over your base of support. Image both the cone head and the long tail extending your body.

Identifying landmarks for alignment
10 minutes

Standing: Find your plumb line. Visualize the cone head and tail. Using your fingers, touch the center of each ear (the small flap that covers the hole). Let the elbows extend horizontally. Imagine lines from each finger that meet in the center of your skull. Do a small "yes" nod from the place where these fingers meet. This is the place where the skull sits on the top of the spine.

○ Move your hands to the sides of your ribs. Place your thumbs behind, fingers in front of your rib cage; feel the oval shape of the ribs, and rotate the cage forward and back. Touch the place on each side of your ribs that feels like the center, or the axis.

○ Place your hands on the sides of your pelvis. Walk down your body until you feel a large knob on each side, at the top of your leg. (It is the knob which touches the floor when you lie on your side; lie down if you can't find it.) This is the greater trochanter of your femur.

○ Bring your hands down the sides of your legs, touching the sides of your knees. Image the center of each joint. Then find the bumps on the outside of each ankle. These are your outer malleoli, the lower end of each fibula. Touch the floor.

○ Bring your fingers around and touch the second toe of each foot (you are bent over now, looking towards the floor, knees bent). Begin drawing two parallel lines in imaginary sand from points twelve inches in front of your feet, through both your second toes, through the center of each ankle and heel, through the center of each knee, through the center of each hip socket (deep in the pelvis). Repeat this movement several times, tracing the parallel alignment of foot to knee to hip in postural alignment. When you are finished, continue your hands from the pelvis up the sides of your ribs, the sides of your skull and up into the air. Repeat the whole process several times to identify the points of alignment.

Balancing the body weights
15 minutes

Standing next to a full-length mirror, with a lateral view (side to mirror) far enough away that you can see your full body. Look straight ahead (not at the mirror). Close your eyes and find plumb line. Then, shift into your most comfortable "hang out" posture; something that feels very natural and familiar.

❍ Rotate your head very slowly to the side to look at your alignment in the mirror. Without judging "good, bad, too fat, too thin, etc.," drop a plumb line in your mind's eye down through the body you see in the mirror. Ideally, a plumb line can pass through all three body weights, through the center of the knees, and the center of the ankles without getting interrupted by sharp angles or passing outside of the body. Try not to readjust yet. Observe any diagonals that pass through the body: skull to rib, rib to pelvis, pelvis to knee. Then return your head to facing front, and begin your work, eyes opened or closed as needed. Beginning at your feet and ankles, bring your feet into parallel alignment below your hip sockets (about four inches apart). The second toe should be facing straight forward like your nose (there are always individual exceptions, so don't force anything that feels painful). Find the malleoli of your ankles. Then touch the greater trochanters of your hips – the knobs on the outside of the legs – and line up the trochanters directly over the ankles. (This might necessitate a major shift of your alignment making you feel as though you are falling backwards or forwards; try to continue your work systematically until you reach the skull.)

❍ Then place your hands on the sides of your ribs, thumb in back and fingers in front. Rotate your ribs forward until they are centered over your ankles and over your greater trochanter. Thus the upward facing bowl of the pelvis matches the downward facing bowl of the ribs and diaphragm, rather than the pelvis and ribs being tipped open, hinged in your back, like an open oyster shell. As you move the ribs forward or backward you affect the relationship of thorax to the skull and pelvis. Check too at this point that the knees are not locked. Keep them relaxed so the energy can flow from your feet to your head (and you can respond to changes in your alignment without falling over).

❍ Then place your hands on the center of the ears to do a "yes" nod. Finish the nod so that your eyes, when open, will be looking straight ahead, as though at someone your own height. The eye is in line with the ear, parallel to the ground. The top of the skull makes another downward facing bowl, parallel with the diaphragm of the thorax.

❍ Tap the top of the head, and, if you are willing, pull a few strands of hair directly up to stimulate your vertical plumb line. *Slowly* rotate your skull around the vertical axis towards the mirror without disturbing your posture. Again, rather than judge, drop a plumb line down through the body parts in the mirror and see how they are "stacked up" in relation to verticality. They should be like three spools strung on a thread, suspended over the center of your feet with your knees relaxed.

❍ Now, you can start the process again, looking in the mirror: ankles to hips, hips to ribs, ribs to skull, skull to plumb line. Be easy on yourself. Tightening muscles in frustration or judgment will *not* help. Posture is a dynamic process. Instead of looking for a fixed position, a pose, or a "right" way to stand, we are establishing a vertical energy line around which the body parts can orient themselves. Return to your hang out posture. Close your eyes. Feel it. Then realign from the feet up. Check the mirror. Repeat. What feels "right" is often what is most familiar; so part of the work is getting comfortable with the changing sensations.

Partnering:

Do this work on your own first. Then partner. Touch each body landmark as it is being aligned; resist forcing or manipulating someone into a position from outside. The shifts are subtle and need to be experienced from the inside to create lasting effect. It is helpful to rotate the skull for your partner in the "yes" nod, then gently lift on the skull to assist the sensation of verticality. Tap the top of the head, and pull the hair on the plumb line.

37

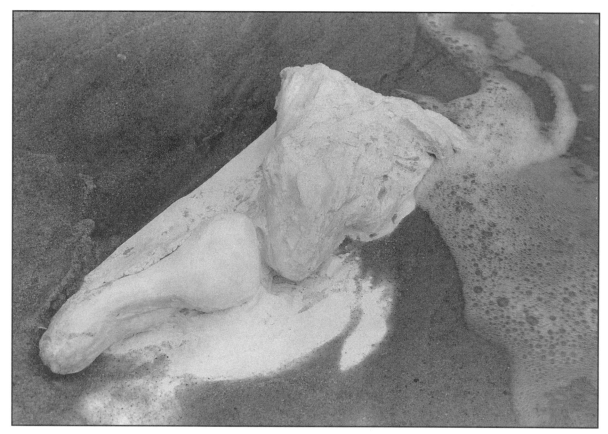

Sculpture: Harriet Brickman
"Beached Forms Series"

THE NATURE OF BONE

Our skeleton is alive. The 206 bones in the human body are living tissue. Both red and white blood cells are produced in the marrow of long bones. The skeletons we see in museums, art classes, or classrooms are the mineral salts that remain after death, representing approximately 65% of bone by weight. The other 35% of living bone is a mixture of connective tissues in a watery ground substance with many blood vessels and nerves. The mineral salts are deposited around the softer tissues, forming a porous inner structure. Bones are dense around the outside and at each end providing lightness at their core as well as considerable strength and resiliency.

Bones are the framework of the body. Bone is covered by a connective tissue sheath, called periosteum on the outside surfaces and endosteum on the inside. This tissue is vital to bone nourishment, growth and maintenance. Because it is a connective tissue, the periosteum and endosteum provide a continuous interweaving with other connective tissues such as the ligaments which attach bone to bone, the tendons which attach muscle to bone, and the joint capsule, which secretes and contains the lubricating synovial fluid. Connective tissue also surrounds every muscle, blood vessel and organ, forming a continuous connective span throughout the body which is woven into the fabric of bone.

Bones change throughout our lifetime. They exist in the fetus as cartilage as early as the fourth or fifth month; ossification (the process of osteoblasts taking calcium and phosphorous from the blood and fixing them into bone crystals) progresses rapidly throughout the first four or five years of life, and continues on until approximately age twenty-five. Bones continually respond to use and abuse. They are affected by evolutionary and genetic heritage, but also by proprioceptive stimulation, balance of the body weights through the skeleton, diet, exercise, trauma, illnesses and injuries, emotional experiences, and life patterns of work and play. Bones need balanced activity to maintain their health: too little use results in atrophy (bone deterioration), too much in fracture. During the aging process, both exercise and sufficient calcium in the diet are necessary to ensure bone strength.

Bones grow, while maintaining their balance of lightness and durability, by a process of calcium deposits on the outer surfaces and calcium dissolution on the inner surfaces. Primary growth occurs between the shaft (diaphysis) and the ends (epiphysis) of long bones. This cartilaginous juncture is called the epiphyseal line (the cartilage plate). New cartilage is

Assembling the Skeleton

When I came to teach Anatomy and Kinesiology in the dance program at Middlebury College, we purchased a human skeleton (from the Guest Artist budget!).* My colleague, Caryn McHose, and I unpacked the skull and spine and began carefully assembling the other bones. As we held each one we said, "Oh, this is the way a femur (thigh bone) looks. Oh, this is the size of the pelvis. Oh, this is the shape of the clavicle (collar bone)." When we got to the top, the screw which connected the skull to the metal support stand was broken. So we disarticulated our skeleton and mailed it back to the company. A month later, our second skeleton arrived. We were surprised that the femur was different, the pelvis had new proportions, the clavicle was less curved. We were reminded of the uniqueness of each person.

---

* The purchase of human skeletons is now illegal in the U.S. As we change to pliable plastic models and are given a more responsive image of bone, we are also given the illusion that all bodies are identical — from a common mold. It is easy to forget that variations in genetic code and the effects of life experiences on bone make each skeleton individual.

Chinese talisman to protect the body, composed of the character for life, sheng. Calligraphy by Chungliang Al Huang.

constantly produced to add length, and the cartilage at the epiphyseal line is turned into bone. The bone then must be hollowed by the dissolution process. The calcium balance in the blood is also monitored by this process; when calcium is needed, the bone is dissolved and when there is excess, it returns to the bone for storage.

Bones have multiple functions in the body: to produce blood cells, to serve as storage sites for calcium and phosphorous necessary for the blood, to protect the organs, to transfer the weight of the body to the earth, to absorb the impact of locomotion, to provide attachment sites for muscles and connective tissues, and to offer a bony framework for moving and balancing the body. ❖

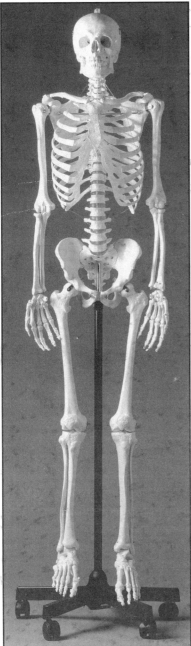

Skeleton: anterior view

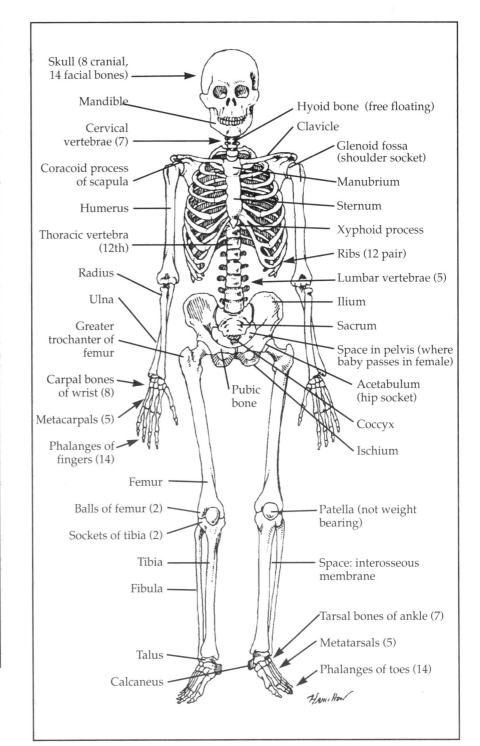

Articulating the bones
15 minutes

Lying in constructive rest: Bring your awareness to the bones. Begin moving from this position to articulate each bone in the body. Play the scales of the body, like a musician, and move each bone in sequence from the feet to the head.

○ Begin with the toes. Continuing with the eyes closed, do a dance of the toes, moving any way that feels good. Let the legs follow the movement, but keep your awareness on the toes.

○ Begin moving your lower legs; allow your lower legs to move in any way that feels good to them. No judgment, no evaluation, just notice the movement of the lower legs.

○ Move from your knees. Let them lead where they want to go. If you need to move fast, or sharp, or through space, open your eyes enough to see objects or other people, but listen to your body. Continue exploring with your knees, bringing your awareness to your upper leg and hip sockets. Let them begin to guide the movement. Explore all the directions a hip joint can move.

○ Follow all the lines and arches and curves of the pelvis. Notice any images, ideas, sensations, emotions as they come, staying focused on the pelvis.

○ Notice how the pelvis moves the spine. Feel all the directions a spine can move: it can twist, it can roll, it can undulate. See what your spine wants to do at this time, on this day, in your body.

○ Connect the spine to the skull and jaw. Let your whole body follow the skull. Allow the jaw to initiate your movement.

○ Begin to hear the shoulder girdle, especially the shoulders. Explore movement or stillness in the shoulders. See what they want to do.

○ Bring your focus to your elbows and lower arms. Explore movement in this area of your body. Allow any other areas to participate.

○ Follow your wrists and hands. Do a dance of the hands and fingers. They often have a lot to say. Try not to judge or analyze your movement. Just follow. See what's there.

○ Move any body part that wants to move. Feel the relationships between body parts. Notice the whole.

Moving from bone
15 - 20 minutes

Lying in constructive rest: Bring your awareness to the bones, then to the layers within the bones. 3 minutes.

○ Move with awareness of the hard, cancellous bone. Feel the solidity and strength of this outer portion of the bone. Begin to travel through space, and to work standing. 3 minutes.

○ Move with your focus on the marrow, the life-giving center of the bones. See if your movement changes. Feel your whole body move from this fluid center. 3 minutes

○ Alternate moving between the two. 3 minutes.

○ Finally, move with awareness of the periosteum, the connective tissue covering of the bone that interweaves with the muscles and other fascia throughout the body. Observe any changes in movement from working with this layer. 3 minutes.

○ Move from the whole body, integrating all the bone layers. 5 min.

✎ Talk or write about your experience.

Drawing: Jim Butler
Untitled

SKULL, JAW, AND HYOID BONE

The skull is the top body weight, balanced effectively on the spine. It is comprised of the eight bones of the cranium to hold and protect the brain and the fourteen bones of the face. The cranium is formed by the frontal, occipital, sphenoid, ethmoid, two temporal, and two parietal bones. (See drawing, page 44) Each cranial bone connects to the next through fibrous suture joints. These interlocking connections, which look like rivers viewed from the air, create stability at the bone level. The sutures are woven together both by their bony shape and by fibrous connective tissue, requiring a minimum of muscle to keep them secure. They are, however, responsive to change, with subtle shifts in relation to inner pressures and exterior blows. They function as effective shock absorbers for the brain while providing a relatively light and efficient protective covering. Tension can cause muscles of the scalp to constrict the sutures, blocking the responsiveness of the skull. Suture massages in hands-on work can relax the muscles and thus alleviate pressure built up within the skull and allow realignment of the plates in relation to each other. The facial bones give structure and form to our face and mouth and are connection sites for many intricate muscles.

The sphenoid bone is butterfly-shaped and lies in the transverse plane. It is the only bone which goes horizontally through the skull, making a

Sphenoid bone

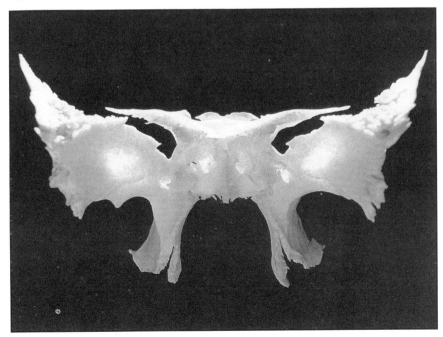

Masks

I was raised in the era of Little Miss Sunbeam, the Jantzen Smile Girl, sororities, cheerleaders, and Miss America pageants. I participated in each of these from childhood to high school to college. The smile was an important asset, useful to monitor communication and response. The many tiny muscles of the face created a mask, with the smile a strange, dull, reflex when disconnected from emotional response, sometimes dazzling as it worked to hide the same. At one such contest, I was wearing white gloves and a polka dot dress and was smiling at the judges in a small, close room. I remember thinking,"What am I doing here?" And I stopped smiling. I felt it dissolve from the inside out. It was the first time that my face didn't work.

❖

When I was teaching Primitive and Ancient Dance History at the University of Utah, I was invited to give a lecture on "Shamanism and the Origins of Perform-ance" at an art museum in the northwest. The lecture was to be followed by a performance of Indian and Eskimo folk tales by myself and two colleagues as part of a major exhibition of Northwest Indian and Eskimo Art. As the audience entered, we sat in a hall filled with dazzling artifacts while painting our faces in front of display cases to match particular masks. I was copying a sky blue shaman's mask, accented with strong lines of black and white. As I painted the last stroke across my cheek, I realized I couldn't speak. At first, I thought it was an interesting idea, "How can I transform into a mask and then give an intellectual talk about shamanism." Then it was real. My mouth was silent.We all went upstairs to the performance area and got into our positions. My colleagues waited for me to stand up and speak. After an uncomfort-able pause, one of them began the story-telling. I never moved from my position, and they gave the performance around me. At the end, we walked downstairs, and retrieved our own faces. Dripping with water, I finally said, "I couldn't talk." I hadn't believed that masks were real.

base for the brain and a ceiling for the mouth. The sphenoid bone also supports the pituitary gland, and is one of the seven bones forming the socket for the eye.

The mandible, or jaw, connects to the skull at sockets in the right and left temporal bones of the skull, slightly anterior to each ear. The temporomandibular joint, or TMJ as it is commonly called, is composed of the curved "ball" of the mandible, the socket in the temporal bone, and a disc to cushion movement surrounded by synovial fluid and a joint capsule. Many ligaments provide stability and direct the angle of muscle pull on this much used joint. In a relaxed state, the jaw would hang open, pulled by gravity. Keeping your mouth shut necessitates constant contraction of the masseter muscles, which can be felt by massaging your jaw and cheeks. The masseter is involved in chewing and is one of the strongest muscles of the body.

The hyoid bone is a delicate free-floating bone on the front of the neck. Shaped like a small horseshoe, it is suspended by ligaments from the skull (from the styloid processes) and stabilized by muscles attaching to the floor of the tongue and to the mandible. You can feel this bone by gently touching below and under the chin. Early developmental patterns and reflexes such as sucking, swallowing and head righting are intimately connected with the hyoid bone, due to muscle attachments. Changes in spinal patterns often involve work with this small but important bone of the axial skeleton. ❖

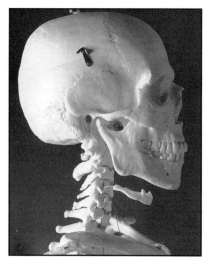

Skull with hyoid bone: right lateral view

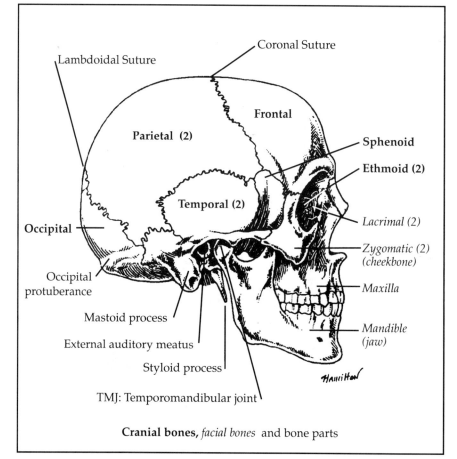

Cranial bones, *facial bones* and bone parts

Suture massage
20 minutes each

In constructive rest, with a diagram of the sutures of the skull beside you: Begin tracing the central suture between the right and left parietal bones on the top of the head. Press gently but firmly with the flat of your fingers.

○ Travel forward and walk your fingers horizontally across the sutures of the frontal bone. Massage the area of the bone connections to feel the small ridges and indentations. Even if you are not specifically feeling a suture, the touch of the fingers will bring sensory awareness to the skull, so don't worry about being exactly right.

○ Return and trace around the suture of the occipital bone (the lambdoidal suture). You may need to lift the skull carefully off the floor.

○ Trace the curved sutures of the temporal bone around the ears, walk your fingers carefully out the zygomatic arch, or cheek bone, and back. Find the TMJ joint with light touch. Feel gently for the suture of the sphenoid bone. Lightly trace the rim of the eye socket.

○ When you have mapped the sutures of the skull, gently massage the scalp. Pull lightly on the hair to stimulate the receptors in the hair follicles.

As you release your fingers from your partner's head, feel the change from touching the bones of the skull, to the muscles and skin of the scalp, to the hair, to feeling just the heat of the head, and notice when your hands return to neutral and you separate your energy from your partner. Place your hands on the floor.

○ Have your partner roll on her/his side, keeping the head in contact with the floor. Slowly come up to seated. Again, slow movement reduces dizziness. Walk around the room. Talk with your partner to compare experiences.

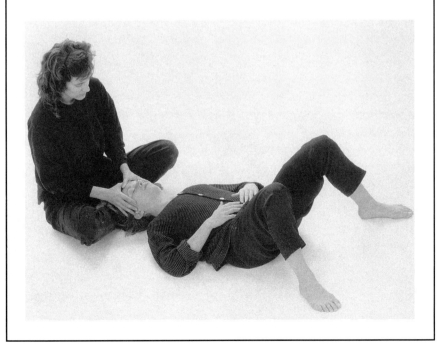

Working alone:

Suture massage can be done on your own. Look at a diagram of the sutures before lying in constructive rest. Feel your own skull, eyes closed, and trace the connections between the bones as outlined at left. Massage the full skull. Roll on your side and come to seated.

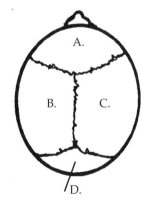

Skull showing sutures, from above:
A. Frontal Bone
B. Left Parietal Bone
C. Right Parietal Bone
D. Occipital Bone

Holding the head
15 minutes

With a partner: The person "touching" should be seated comfortably on the floor, balancing the body weights over the pelvis for efficient alignment. The less tension you have in your own body, the more your proprioceptors are free to feel your partner. (If you are tight, all you feel is you!) Breathe fully. The person being touched should lie in constructive rest, eyes closed, with the head towards the seated person, and the focus on breathing.

❍ Place your hands very slowly under the neck of your partner. Lift the skull about two inches off the floor, holding it securely in both hands. The slower you work, the more relaxation can occur.

❍ Begin a small "yes" nod. Feel the skull move on the first vertebra, the atlas. Feel the skull move the neck.

❍ Rotate the skull around its vertical axis, turning it slowly in a "no" action to the left and back to center; to the right and back to center. Encourage your partner to release the full weight (13-20 pounds!) of the skull by the careful support of your hands.

❍ If you are being touched, focus on your breathing. Image that your head is lying on a pillow, or that you are lying on the beach in the sand. Feel the full weight of your head as it is lifted. Let your partner do the work. As your head is rotated, image beans in a beanbag slowly rolling from side to side, or water in a plastic bag shifting in its container as it is moved, or that the head is being rocked by a moveable pillow platform. Let your eyes float in pools of water in the sockets. Relax the hold on your vision under closed eyelids. Give feedback to your partner if something is uncomfortable or if you don't feel safe; keep talk to a minimum.

❍ Continue to slowly circle the skull any direction, feeling the shifts of weight under your hands. You may feel muscle holding, or blood pulsing, or nerve impulses such as small jerks. Keep the movement small and slow to facilitate maximum release. Repeat a small "yes" nod with your partner; a small "no." When you are ready, place the skull back on the floor so slowly that your partner doesn't perceive the change between the surface of your hands and the surface of the floor. Touch the floor with your hands to separate your energy from your partner.

❍ Have your partner slowly roll onto her/his side and up to seated. Quick movements may induce dizziness. Then stand and walk to find neutral. Talk to each other about your experience. Remember, it is not about judging good or bad. Speak about what happened.

❍ Reverse roles and repeat.

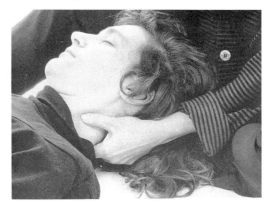

Proximal and distal articulation of the TMJ joints

5 minutes

Seated or standing: Place your fingers on the joints between the skull and the jaw, the temporomandibular joints. These can be found on each side of the skull slightly in front of the ears.

❍ Gently open and close your jaw and feel the movement of the joints under your fingers. These are delicate connections, so don't press too hard.

❍ Place your elbows on a table or on the floor. Grasp your chin with both hands to stabilize the jaw. Move the skull while the jaw remains still. This process will move the head up and back in space, away from the chin. The sockets of the temporal bones travel, or excursion, around the stable balls of the jaw bone (mandible).

❍ Stabilize the skull and move the jaw. Stabilize the jaw and move the skull.

❍ Then place your fingers once again on the joints and move the skull and the jaw equally around the discs. This equal and opposite (balanced) motion helps to reduce tension in the jaw and neck.

This excercise can be useful for individuals with TMJ syndrome to help center the discs in the joints.

Skull moves back, jaw moves down, both move

Photograph: Bill Arnold
"Steps, California"

AXIAL SKELETON:
Skull to Spine

The skull and spine are part of the axial skeleton; this is our ancient fish skeleton before weight bearing appendages were added. The axial skeleton includes the skull, the spine, the "rib cage," and the free-floating hyoid bone. More specifically, the spine involves a mobile column of vertebrae: seven cervical, twelve thoracic, five lumbar, the five fused sacral vertebrae, and the four fused coccygeal vertebrae – our ancient tail. The "rib cage" involves twelve pairs of mobile ribs, plus the sternum, manubrium, and xyphoid process. The free-floating hyoid bone is suspended on the front of the neck from the skull. Note that the sacrum and coccyx are part of the axial skeleton, but the pelvic girdle is part of the appendicular skeleton, formed when appendages became weight-bearing.

The weight transfer from the skull to the spine is key to postural alignment. The head weighs between 13-20 pounds in the average adult. This weight is passed to the rest of the spine through the first vertebra, the atlas. It travels down the bodies of the vertebrae, spreads around the bones of the pelvis through the sacrum, and is transferred to the ground through the legs and feet. The spine can be seen as an extended pyramid with the cervical vertebrae at the top. Each vertebra gradually increases in size so that the larger lumbar and sacral vertebrae form a stable base to spread the weight into the pelvis. The weight transfer from the skull to the spine is crucial for spinal alignment and efficiency of the nervous system. The spinal cord hangs like a tail from the brain, enclosed in the space made by the spiny projections to the side and back of each vertebra. These are called the transverse processes (laterally, one on each side) and the spinous process (the sequential bumps down the back which we feel in back rubs). The skull sits on the atlas, transferring its weight anterior to the cord. Two bony protuberances, called condyles, of the occipital bone form the "feet" of the skull and connect to the receiving sockets of the atlas. Balance at this junction keeps the transition from brain to cord free from pinching or pressure. The foramen magnum, a hole also formed by the occipital bone at the base of the skull, allows the cord to pass from the brain down the spine behind the weight-bearing bodies of the vertebrae. The spinal cord begins at the base of the skull and ends between the first and second lumbar vertebrae; spinal taps are taken below this area to avoid the cord. There it branches into nerves forming the "horse's tail," and spreads through holes in the sacrum to innervate the legs. Within the cranium and the vertebrae, the brain and spinal cord are bathed in cerebrospinal fluid for nourishment and protection.

On Alert

I have a close friend who is a Vietnam veteran. He described to me how he often chose to "walk point" when he was in the field. This lead position for a foot soldier was a likely place to be killed or injured by a mine or by a surprise attack. He worried that someone would get killed, so he preferred to be in the position himself. His spine never relaxed in nine months of active duty. Now he has severe spasms in his back. His work with Vet groups and in therapy has begun to release the tensions from the experiences of war. Yet even a small disturbance can result in a feeling of being on guard and a rigid spine. The more his capacity for relaxation increases, the more extreme the spasms. He is feeling the pain.

❖

A student, whose father is alcoholic, suffered daily from migraine headaches and severe back strain. Often, while he was smiling or talking, his forehead would be scrunched forward with a furrowed brow. One day he asked me to work on his back. As I touched his spine, he started crying. I could feel the amount of tension stored in his strong body to keep him upright, to keep him from collapse. His father was a respected doctor, but was also an alcoholic. He would treat his son lovingly one moment, and then would be irrationally angry a short while later. The student had learned never to let down his guard. If he relaxed and responded to the warmth, he was that much more vulnerable to attack. Now, he wanted to let go; he was tired of holding on.

The atlas is the only vertebra which can move independently from the rest of the spine. The skull can rock forward and backward around its horizontal axis making a "yes" nod. It rocks on its condyles in the two sockets of the atlas, and the skull and atlas together pivot around the second cervical vertebra, the axis. The axis has a vertical projection called the odontoid process, around which the atlas sits, like a finger in a ring. This allows movement in the transverse plane, such as a small "no" rotation. Feel the "yes" nod to locate the atlas, the "no" nod to locate the axis. Movement of the axis and every other vertebra, however, results in collective action of the spine. Activity of any one vertebra will lever into the rest of the column through the transverse processes to ensure safety of the cord. Stress and angles of force should be distributed throughout the spine. Like a snake, movement of the spine travels collectively and rhythmically through the whole structure.

SPINAL CURVES

Our spine is a progression of alternating curves. If we return to our diagram of postural alignment, we can draw the spinal curves which connect the three body weights. There is an anterior curve for the seven cervical vertebrae forming the neck, a posterior curve for the twelve thoracic vertebrae forming the back of the rib cage, an anterior curve for the five lumbar vertebrae known as the lower back, a posterior curve for the sacrum, and the beginning of an anterior curve for the coccyx, or ancient tail. Each of the anterior curves touches the plumb line. Notice the pairing of spinal curves: cervical and lumbar as anterior curves, thoracic and sacral as posterior curves. A problem in one area is often reflected in the other; for example, tension in the neck may result in pain in the lower back. As we learn about the spine, draw its curves, see it reflected in images around us, and tell stories about our own relationship to verticality, we can better understand the importance of this responsive central axis of our body.

In postural alignment, balance of the spine lives in the mobility of its opposing curves. Our goal is to have a responsive spine, not a "straight" spine. The function of the spinal curves is for shock absorption; any impact to one portion is absorbed in the whole structure, protecting the cord, brain, and vital organs. The structure of the spine allows three-dimensional mobility so that we can bend forward, lean sideways, and twist around behind ourselves, through the three planes of movement. Simultaneously, at the bone level, it offers postural stability through the bodies of the vertebrae, and a stable latticework for

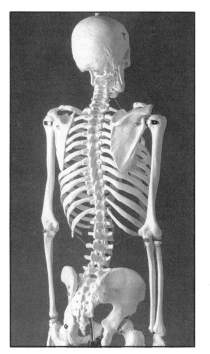

Spinal curves and three body weights

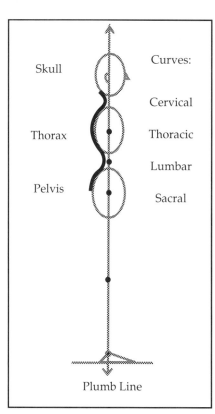

Skull

Thorax

Pelvis

Curves:

Cervical

Thoracic

Lumbar

Sacral

Plumb Line

50

attachment of appendages, muscles, and fascia. It is this balance of mobility and stability which we seek in our spine.

The top of the plumb line extends the vertical axis towards the sky, to lightness. The bottom of the plumb line reaches to the earth, towards the pull of gravity. In postural alignment, centrifugal force and gravity work as allies. It is the dialectic between compression and elongation which energizes our body and gives span and tone to the tissues. Any change in one body part affects the alignment of the whole. Movement of the tissues affects the bones, movement of the bones affects the tissues. Further, if the spaces between the bones are aligned, the body will be aligned.

Movement is key to balance. The body is never still. Neurological activity is a sign of life; cessation of neurological activity, a sign of death. We rely on nerve reflex arcs in the lower leg for constant contraction, release, and recontraction of muscles to keep us in dynamic balance. As we know from standing in lines, it takes more energy to hold still than to move. Directives such as "Hold still," "Stand up straight," or "Suck in your gut" encourage gripping and tension and restrict our stability and mobility. If you are pushed when you are tight, you might easily be pushed off your base. If you are pushed when you are relaxed, the response is one of subtle adaptation.

In efficient body use, the weight is transferred to the earth through the skeleton; the fascia and connective tissue provide basic stability and span at the joints; muscles are free to move the bones – provide mobility; the nervous and endocrine systems are unhindered and able to direct, interpret, respond; the fluids travel to all parts of the body to bring nutrients, bathe, clean, connect and provide oxygen; and the organs provide vital functioning with minimum compression and restriction. Ideal postural alignment is dynamic, everchanging, and interactive with internal and external stimuli. ❖

❖

A cheetah strolled across the plains in the Masai Mara game reserve in Kenya, ready for a kill. As she walked, her presence was telegraphed to all of the animals on the horizon by alert spines or flicks of tails. Her main prey, the gazelles, followed along with her as she traveled, their muscles ready to respond. They preferred to keep their predator in sight, rather than be surprised by an attack.

❖

The Masai men and women in Africa probably have the most elegant spines in the world. Many photographers and writers have tried to capture their vertical presence on the land. Walking for miles across the expanse of plains, or standing amidst a herd of cattle, their tall, lean bodies are studies in ease and beauty. Their houses, in contrast, are low, with tiny doors and few windows. To go inside, one has to bend low, and maneuver carefully through small passageways. Once inside, you remain folded towards the ground, several people to a room, bathed in smoke and warmth from a small fire. It feels like a womb: dark, nurturing, enclosed. I was struck by the balance between compression and extension. Their bodies show the confidence which comes from knowing boundary and boundlessness.

Walking down the spine
20 minutes

Seated, body weights balanced over the pelvis, eyes closed: Feel your vertical plumb line. Do a small "yes" nod with your skull to find the connection between the skull and the atlas. Begin rolling down the spine vertebra by vertebra. Image you are walking down a ladder, curving from each vertebra as you step. Use your fingers to touch the vertebra being moved.

○ Touch the base of the skull. You will feel an indentation. This is the atlas which has no spinous process. Roll the head forward from the skull and atlas. Breathe deeply, keeping awareness on your fingers.

○ Walk your fingers down to the second cervical vertebra, the first "bump" on the neck. Roll the head forward from the axis, the second cervical vertebra. Breathe deeply, feeling the movement under your fingers. Continue down the pointed protuberance of each of the 7 cervical vertebrae of the neck. If you are unclear, approximate and keep going; you are still bringing sensory awareness to the area.

○ Touch the first thoracic vertebra. Either the seventh cervical or the first thoracic vertebra are generally the largest vertebra at the base of the neck. Curve the spine forward from this vertebra; breathe.

○ Release your hand if it is too difficult to reach your arm around your back. Continue to imagine the touch and locate the second thoracic vertebra; release the spine forward from this place. Breathe. Continue down the 12 vertebrae of the rib cage, one at a time.

○ Using your fingers, as your head and ribs are curved forward, feel for the bottom rib and walk your fingers towards the spine. The vertebra which attaches to the bottom ribs is the twelfth thoracic vertebra. Many important muscle attachments occur at this vertebra. Breathe into your fingers.

○ Continue moving your fingers to the first lumbar vertebra, below the rib cage. The spinous processes of the lumbar vertebrae are slightly larger and flatter. Hang forward from the first lumbar vertebra and breathe.

○ Walk down the five lumbar vertebrae, one at a time, allowing the spine to curve forward from each one.

○ Now you have arrived at the sacrum, the five fused vertebrae which form the keystone of the pelvis. Place your hands in the center of the pelvis and feel this triangle of bones as you breathe. Use your fingers to touch the coccyx, the base of the spine, your ancient tail. Your spine is curved fully forward from your tail to your skull, like the ancient symbol of the snake eating its tail. Feel the curve.

○ Return the coccyx and sacrum to plumb line. Feel the weight drop down to come up. Begin counting up the vertebrae, moving each back towards plumb line as you walk up the spine: 5th lumbar, 4th lumbar, 3rd lumbar, 2nd lumbar, 1st lumbar, 12th thoracic, 11th thoracic, 10th thoracic, 9th thoracic, etc. up the seven cervical vertebrae and then bring the skull back to plumb line. Tap the top of the skull and image the cone head extending the vertical axis.

Rolling down the spine
5 minutes

○ Standing: Begin with the "yes" nod and then let the skull fall forward to initiate the curve. When you are upside down, relax the neck and spine into gravity. Hang from your tail and bend the knees slightly for support. Bring the tail down to roll up, returning to vertical plumb line. This counterbalance between pelvis and head is like a bobbing duck, alternating head and tail, or like a counter-weight at a well pulled down to bring the bucket up, released to drop the bucket down.

○ Repeat the bobbing action, alternating the head falling, the tail falling. Go slowly enough to feel the articulation of each vertebra. Remember, movement of the spine is collective action: even though we are articulating each vertebra individually, they each affect the next. We seek equal movement from each vertebra so there is no stress in any one area.

Walking down the spine: with a partner
20 minutes

This exercise is very effective because of the increased sensory stimulation through touch. Begin with one partner seated on the floor behind the other. The person in front has eyes closed, and is seated on plumb line. The person doing the touching is also seated efficiently for weight transfer. Keep a secure base of support as you touch someone else. Image the place being touched in your body.

○ Place your hands on each side of your partner's head, thumbs at the base of the skull. Keep your fingers together so your little finger does not go near the eye. Lift the skull slightly on the vertical axis.

○ Place your finger at the base of the skull, in the indentation, and ask your partner to release the head forward from the first cervical vertebra. Be touch specific, so that you draw the focus where you want it. (Avoid generalized rubbing, or random movement of the other hand.) Continue walking down the spine as described in "Walking down the spine," touching the bony protuberance of each vertebra. Don't worry if you can't feel it clearly. The vertebrae are approximately an inch apart. Continue to walk down the spine and look for the major landmarks: The large vertebra at the base of the neck which is the first thoracic vertebra, the vertebra attached at the bottom of the ribs which is the twelfth thoracic vertebra, and the junction of the spine to the sacrum, which is between the fifth lumbar and the sacrum.

○ At the sacrum, place your hand, fingers down on the triangle of the sacrum. Encourage your partner to breathe into this area. Then, "itsy, bitsy spider" your fingers back up the spine, saying the name of each vertebra as you touch, beginning with the fifth lumbar. Give your partner time to bring each vertebra back towards plumb line before moving on. When you reach the skull, again place the hands carefully on the sides of the head and give a gentle lift. Tap the top of the skull and ask your partner to open her/his eyes. Change partners. Stand and do the "Rolling down the spine" exercise described above.

53

Rocking the spine
5 minutes

Lying on the floor on your back, legs extended: Begin gently rocking your heels on the floor.

❍ Relax your spine, and allow the movement to travel up your legs and sequentially articulate the vertebrae up through your skull.

❍ Do this movement easily, like moving in a rocking chair, until the body tissues relax around your bones.

Three-dimensional spinal patterns
10 minutes

Standing, legs comfortably apart: Begin to swing your pelvis forward and back between your legs. Think of swinging it from your tail.

❍ Allow the movement of the pelvis to become smaller and undulate your spine from the tail to your head. Keep your knees bent.

❍ Repeat until this movement becomes very small and rhythmic. This is the movement a whale makes as it propels itself through water; imagine its head on the top of your skull and a flat tail below.

❍ Swing the pelvis side to side. Initiating with the tail (sacrum and coccyx and imaginary tail), establish a small lateral undulation through the spine with this side to side motion. This is the movement a trout makes when it swims through water; imagine its head on the top of your head and a vertical tail below.

❍ Begin rotating the pelvis, spiraling the spine around the vertical axis. Feel the rotation travel up the spine; image a flag wrapping around a flag pole; reverse directions. Notice that the pelvis gives an equal and opposite force for the rotation of the spine and head. This is the movement an otter does to roll over in the water. Rotate smaller and faster; feel the subtle, sequential spiralings around the spine like stripes on a barber shop pole.

Caring for your spine:

• Learn the structure of your spine, visualize it clearly. (pp. 49, 55) Distribute weight equally through the bodies of the vertebrae. (p. 35)

• Check skull, rib, pelvis alignment. (p. 15) Know that the spine houses and protects the spinal cord. Relate pain or ease in the spine to alignment of the whole body.

• Develop equal and opposite stretch and strength in muscles. (Example, if you repeatedly bend forward, relax by stretching the spine in the opposite direction. Do not force!) Massage the back, hip and thigh muscles to reduce pull on joints. Use sit-backs to identify muscle sequencing for postural support. (p. 89)

• Walk with a relaxed swing of arms and legs (30 minutes daily) to pump fluids and stimulate three-dimensional mobility. Feel your breathing spot as you move, stand or sit. (p. 21)

• Identify movement or thought patterns that encourage holding in the spine/relaxation of the spine. (p. 121) Relax with constructive rest before sleeping. (p. 13)

THORAX:
The Vertebrae and Ribs

The spine consists of seven cervical, twelve thoracic, five lumbar, five fused sacral and four fused coccygeal vertebrae. The atlas and the axis are the connection from spine to skull – the first two cervical vertebrae, and the coccyx is the ancient tail. There is a cartilaginous fluid-filled disc between the bodies of each of the cervical, thoracic and lumbar vertebrae. The function of the discs is to cushion the spine.

Exposure

*On a teaching sabbatical I decided I would be rolfed. I was using the text **Rolfing, The Integration of Human Structure** in my Anatomy course, and I was in Boulder, Colorado for the summer, home of the Rolf Institute. I needed help with my ribs. In every video or photograph of my dancing, my ribs were extended forward. I couldn't feel this alignment, but it was visible. Even with extensive efforts in dance technique classes and bodywork, I was unable to change the pattern. In the Rolfing sessions, we progressed systematically through the whole body working with the fascia. On the day which focused on the ribs, my Rolfer began to press on the area around the sternum. I am very used to touch and was surprised how deeply she was working. But I gave no response. Finally, I moved my body a bit and she released her pressure. As we talked, she said, "I was trying to get you to reflex into protecting yourself. Most people would have responded long before you did. What does it take before you say no?" I trusted her, and knew this work was particular to me, not part of the Rolfing training. So I listened to her question with attentive silence. What did she mean? Would I, in fact, let someone hurt me before saying stop?*

**Costume design: Kristen Kagan
Choreography: Peter Schmitz for dancer Susan Prins**

A vertebra consists of a body, a spinous process, two transverse processes, and a foramen (hole) for the spinal cord. The atlas has no spinous process, as it would interfere with backwards movement of the skull in the "yes" nod.

A simplified drawing of a lumbar vertebra from above would look like:

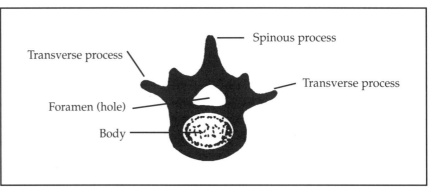

A simplified drawing of a thoracic vertebra and its rib attachment seen from above would look like:

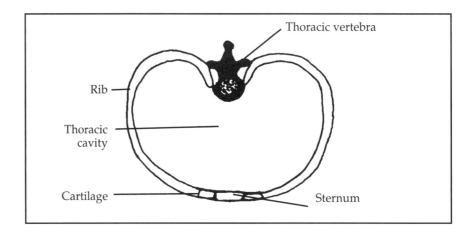

A right lateral view of cervical vertebrae and discs would look like:

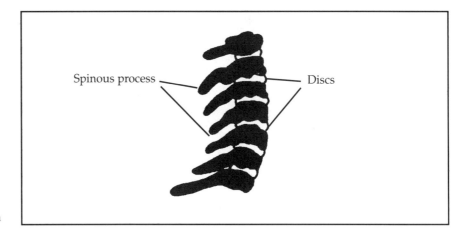

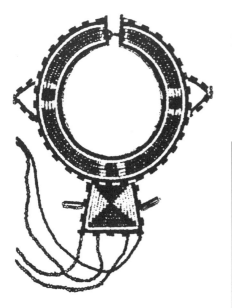

One of many beaded neckplates worn by Masai women in Kenya, East Africa

Collectively, the **bodies of the vertebrae** and their discs transfer the weight of the skull to the pelvis. Note how far they extend into the body cavity. They also serve as attachment sites for the primary **prevertebral** postural muscles (longus capitus, longus colli, iliopsoas and the crus of the diaphragm). The thoracic vertebral bodies are simultaneously the front of the spine and the back of the cavity for the heart and lungs. The **spinous process** protects the cord from behind. It forms the surface of the spine which touches the floor in constructive rest and is the attachment site for many muscles. The transverse processes reach laterally, and protect the cord from each side. They create passageways for the spinal nerves to pass from the cord to the periphery. They also provide attachment sites for the posterior muscles of the spine (erector spinae group). The **foramen**, the hole for the spinal cord, is created by the transverse and spinous processes and is posterior to the weight transfer through the sturdy bodies of the vertebrae. Thus postural support –weight transfer– is anterior to the cord.

The vertebral column is supported by four layers of longitudinal ligaments which integrate the length and depth of the spine. From the front of the spine (inside the body cavity) to the back there is: the anterior longitudinal ligament along the front of the bodies of the vertebrae, the posterior longitudinal ligament on the posterior surface of the bodies of the vertebrae within the vertebral canal (the hole for the spinal cord), the ligamentum flavum on the posterior surface of the vertebral canal, and the supra spinae ligament which connects the posterior surface of the spine, along the spinous processes. Thus, there are four layers of depth for spinal integration. Any curve of the spine should utilize all four layers of support to protect the important spinal cord and nerves. Vertebral alignment, with weight passing anterior to the cord, is key to efficient functioning of the spinal nerves.

The thorax, or rib cage, assists in breathing and protects the lungs and heart. A highly mobile structure, the thorax consists of twelve ribs on each side attached to the vertebrae in the back and the sternum in the front at moveable joints. Seven "true" ribs attach through individual cartilages at the manubrium and sternum on the front surface of the body. Of the five "false" ribs, three attach to the sternum by connecting to the costal cartilage of the seventh rib. The bottom two are considered floating ribs because they have no attachment but end, instead, in the abdominal wall. Each rib forms a hinge joint with the spine by connecting with the body and transverse process of one vertebra, and the body of the vertebra above. Thus the rib can lever into the spinal column for rotation, as well as have up and down mobility in breathing. The floor of the thorax is the muscular thoracic diaphragm which attaches around the bottom of the lower six ribs, and to the twelfth thoracic vertebra. ❖

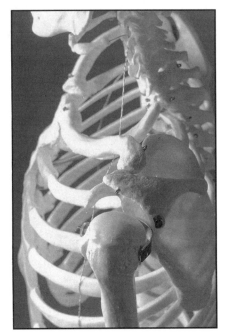

Thorax: left lateral view

Drawing the vertebrae and ribs

✎ Draw: a simplified lumbar vertebra seen from above. Label the spinous process, transverse processes, body, hole for spinal cord and the view being shown. Make an X where the weight passes through the vertebra. Be sure the hole is posterior to the body, between the spinous and transverse processes.

✎ Draw: a simplified diagram of the vertebra and its rib attachments from above. Show how the rib connects to the body of the vertebra and the transverse processes of the spine, as well as with the sternum in front. Note that each rib actually has three joint articulations with the vertebra: with the body and transverse process of one thoracic vertebra, and also with the body of the vertebra above (not visible in overhead drawing). Thus the rib "steers" the vertebrae.

Image the body of a vertebra as the seat of a motorcycle

The transverse processes and ribs are the handlebars. You sit on the body and steer with the handles. Imagine this in your own body – you sit inside the rib cage facing your spine. Place the weight on the bodies, and your hands on the ribs. Steer the body from inside.

Feel the weight transfer down the bodies of the vertebrae
15 minutes

Standing: Find your vertical alignment. Let the weight of the plumb line fall to the posterior surface of the vertebrae -- the transverse and spinous processes. You are "standing in your cord."
❍ Move the plumb forward slightly to transfer the weight to the bodies of the vertebrae. Feel the difference. Try this several times to become familiar with relating the plumb line to the bodies of the vertebrae. Note the sensation of transferring the weight through the bodies, leaving the cord free to hang like a tail from the brain to the lumbar spine and out the peripheral nerves to the legs.
❍ Standing: Move the body from your ribs, move the ribs by "steering" the spinal column.

Tracing the ribs, working alone
15 minutes

Lying in constructive rest: Place your hands on the bottom of your ribs. Feel them move as you breathe. Locate the ends of your two floating ribs on each side.
❍ Using the flat of your thumbs, continue to trace the space between each of your next ten pairs of ribs. Start at the side of the body and massage forward to their attachment at the sternum.
❍ Walk your finger back to the side and feel the diagonal on the back surface as the ribs angle towards the vertebrae. You may feel soreness in the spaces between the ribs; encourage even and equal space between the bones. Be gentle but firm in your touch to avoid "ticklish" sensations.
❍ When you are finished, place your palms on your sternum. Image the depth of your ribcage. Roll on your side. Feel the effects of gravity in this position. Gently rock your ribs. Be aware of the movement of the bones; be aware of the movement of the organs inside the bones.
❍ Roll over to your belly. Feel the ribs move in this position. Continue rolling to stimulate the proprioceptors. Roll feeling the bone; roll feeling the lungs and heart.
❍ Bring yourself to seated. Breathe.
❍ Standing: Turn your body feeling your vertebrae. Feeling your ribs. Feeling your organs.

Tracing the ribs, with a partner
20 minutes each partner

The person being touched: Lying comfortably, belly on the ground, arms extended side or supporting head.

The person touching: Sitting by their partner's left side.

○ Begin "walking down the spine," touching the indentation for the atlas, and the spinous processes of the other six cervical vertebrae. Find the first thoracic vertebra (T 1), which is usually slightly larger than the cervical. Locate the first rib on each side: they originate on the transverse processes of T 1 and make a necklace-like circle to insert below the clavicle on the front surface of the body.

○ Trace the first ribs from the vertebra to the front surface if possible. Return your hands to T 2.

○ Now, trace the space between the first and second rib on the left side of your partner (the side closest to you). Press firmly, but use the flat of your fingers to avoid sharp touch; you can only follow the second rib a short distance before it disappears under musculature (remember, the ribs create smaller circles at the top and get progressively wider in circumference as they descend).

○ Continue down the twelve thoracic vertebrae: Find the spinous process, then the transverse process (the "gutter" along the side of the spine), then the rib. Trace the space between the ribs to stimulate the maximum sensory awareness.

○ When you get to T 7, have your partner roll on his/her side, exposing the ribs you are touching. Continue down the spine, tracing the ribs all the way around when possible to their attachment on the sternum. Feel how the ribs make a strong, downward diagonal on the back surface, then curve up to meet the sternum on the front. Find the floating ribs at T 11 and T 12.

○ When you are finished, place one palm on the sternum, one palm on the center of the back. This helps your partner to feel the depth of the ribs. Then, gently rock the rib cage between your hands. Move slowly. Move the bones. Then move imagining the lungs and the heart; the container and the contents.

○ Have your partner push into his/her hands and arms to come to seated.

○ Stand and walk around.

○ Repeat on the other side.

Tracing the ribs is particularly useful for anyone who has problems with breathing. It helps to bring awareness to the thorax and relaxation to the muscles involved.

BREATHING

Breathing is essential to human life. The heart and lungs are surrounded by moveable ribs which articulate with the sternum in front and the bodies and transverse processes of each of the thoracic vertebrae in back. The sternum, or "blade," protects the heart from impact. The moveable ribs encompass the lungs and assist in the breathing process. The container is responsive to the contents, the contents are affected by the container. Two layers of intercostal muscles interweave between each rib, and facilitate breathing by lifting the lower ribs to expand the volume inside the thoracic cavity for the inhale, and drawing the ribs together to

Sigh

As a baby, I would hold my breath, turn blue and faint. This occured from six months to two years of age, and once – as the family story goes – necessitated a shot of adrenalin in the heart to induce revival.

❖

Midway through my first college faculty position, I developed a condition where I couldn't breathe. I would go to a movie, see an emotional image, and not be able to catch my breath. I was given mild muscle relaxants and told to rest.

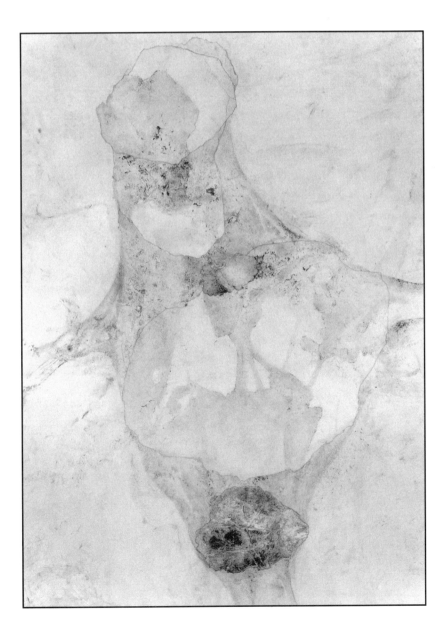

**Collage: Rosalyn Driscoll
"Pilgrimage"**

reduce the volume on the exhale. Holding in the intercostal muscles results in rigidity of the rib cage and a lack of mobility and responsiveness to inner and outer change. These muscles are responsible for only ten per cent of the action of breathing, however. The diaphragm is the primary initiator of breath.

The diaphragm is a dome-shaped tendinous sheath which arcs up into the thorax like a giant mushroom cap. It is made of both contractile muscular fibers and resistant, tendinous connective tissue so that it is both responsive to stretch and firm in its shape. The diaphragm attaches around the bottom circumference of the lower six ribs, connecting spine, ribs, and xyphoid process of the sternum through the transverse plane of the body with an airtight seal. The crus muscle is an outgrowth of the diaphragm and attaches in two portions to the right and left anterior surfaces of the bodies of each of the lumbar vertebrae. The two crura, like the stem of the mushroom, connect the breathing down the spinal column, integrating upper and lower portions of the axial skeleton. The diaphragm is the floor for the lungs and heart, and the ceiling for the liver, stomach, and spleen.

When a muscle contracts, it shortens towards its stable end. As the diaphragm and crus muscles contract, the moveable diaphragm is pulled towards the stable muscle attachments on the lumbar spine. Thus, as the

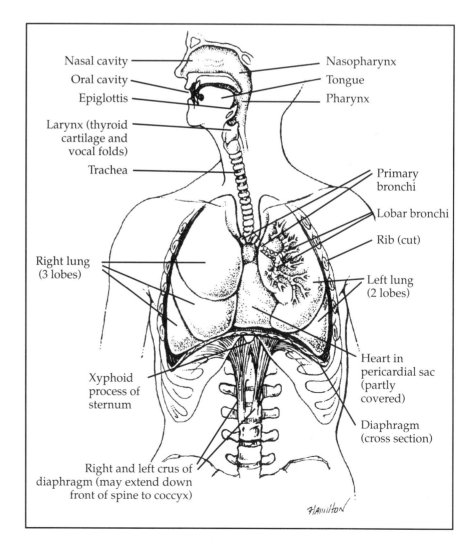

Organs of respiration: anterior view, head rotated

diaphragm descends towards the pelvis (and the ribs fan open laterally due to the contraction of the intercostal muscles), the space is enlarged causing a decrease in air pressure in the thoracic cavity. Air rushes in to fill the vacuum, expanding the lungs and balancing the interior-exterior pressure, resulting in the "in breath." When the diaphragm and crus muscles release, the diaphragm returns to its dome-shaped position and expels the air, causing an "out breath." This constant exchange of interior and exterior space happens reflexively without our attention; it can also be observed and affected by conscious awareness.

As air enters the body through the nose and mouth, it passes through the trachea down the front of the neck, bifurcates into two primary bronchi behind the manubrium of the sternum, which branch again into lobar bronchi, one for each of the five lobes of the lungs. The bronchi divide into bronchioles and again into alveoli, decreasing in size and increasing the amount of surface for exchange with blood vessels. The respiratory bronchiole, alveolar duct and sac, and alveoli constitute a respiratory unit. The oxygen from the lungs is absorbed into the blood vessels, and carbon dioxide is released into the lungs. Three-fifths of the lung volume is blood and blood vessels in route to the heart; there the oxygen is pumped throughout the body.* Cellular breathing is the absorption of this oxygen and removal of waste products in every cell.

The heart is massaged by the movement of the lungs and diaphragm. The thoracic cavity is filled by three lobes of the lungs on the right and two on the left, with the heart nestled into the center and slightly left to balance the symmetry. Holding in any part of the thoracic cavity affects the flow of blood and oxygen in the body. Any training technique or fashion which encourages rigidity or squeezing of the abdomen, ribs, or waist, and restricts full breathing is antithetical to a healthy body. ❖

❖

In Vermont, I observed Caryn McHose teaching a class. She had the participants place their hands on their lower backs, curved in the fetal position, to feel the movement of the muscles under their hands as they breathed. She called it the "breathing spot" and said that the whole spine should move when we breathe — all the way to the coccyx. I saw it in her body. One morning I woke up feeling my spine moving. It was undulating gently with my breathing, like a river inside me.

❖

I began to sing. Your ribs don't move, my teacher observed and placed her hands on my sternum. Do you think of your ribs as a cage? Instead, think of them as bellows, gills, fanning out as you breathe.

❖

I lay still, for hours each day, listening. Each cell breathes, I am told. I cough. Somewhere deep inside the breath knows.

---

* For further information on the process of breathing, see Elson and Kapit's, *The Anatomy Coloring Book.*

Breathing with the five lobes of the lungs
20 minutes

With a partner or alone:
Lying on your belly or in constructive rest:
○ Place one hand on the top right portion of the ribs; feel the bones; then feel under the bones to the movement of the top right lobe of the lung. Breathe into the warmth of the hand. Focus all your awareness on this part of the body. (You will still be using other lobes to breathe.)
○ Move your hand to the middle of the ribs on the right side of the body. Breathe into the warmth of the hand. Focus all your awareness on this middle lobe on the right side.
○ When you are ready, move your hand to the bottom right side of the ribs. Breathe into the bottom right lobe of the lungs.
○ Change to the left side of the body; there are two lobes only due to the placement of the heart slightly to the left of center.
○ Place your hand on the area of the top, left lobe of the lung; remember that the ribs fan diagonally from the top. Breathe into this lobe.
○ Move your hand to the bottom lobe on the left side. Breathe three-dimensionally into this lobe.
○ When you are finished, place your hand on the heart, slightly left of the sternum. Breathe into all five lobes and feel them massage the heart area.
○ Bring your hands to the bottom of the rib cage, and trace where the diaphragm attaches, from the xyphoid process of the sternum around to the back to the twelfth thoracic vertebra.
○ Roll onto one side.

If partnering, place one hand on the sternum and one on the back of the heart, palms flat to the body, seated close to your partner. Gently rock the ribs forward and back. Rock from the bones; then rock from the contents – the lungs and the heart.

Working alone, lying on the floor: begin to roll the body on the floor, eyes still closed. Initiate with the skeleton and feel the ribs as you roll. Initiate with the organs, and feel the lungs and heart as you roll. Notice how the movement is different.

SHOULDER GIRDLE

The shoulder girdle helps us reach, hold, and push away. It is a series of bones wrapped around but separate from our rib cage. The **clavicle**, at its joint with the manubrium, begins the wrapping. The clavicle is an S-curved bone which extends laterally and meets the acromion process of the scapula at about the place a strap crosses your shoulder, the acromioclavicular joint. The **acromion process** curves around to the back surface of the body, like an epaulet on a uniform. There the acromion becomes the **spine of the scapula** and travels towards the vertebrae, horizontal to the earth. The vertical edge parallel to your spine is called the **medial border**, the bottom-most tip is the **inferior angle**, and diagonal up to your armpit is called the **lateral border**. Collectively they form the familiar wing-like triangle of the **scapula** on your back; the whole scapula slides freely over the ribs. The lateral border continues to the side of the body and helps form the **glenoid fossa** or shoulder socket, underneath the curving acromion, like a cave under a cliff. The arm attaches to the shoulder girdle at the glenoid fossa and includes the **humerus** (upper arm), the **radius and ulna** (forearm), the eight **carpal bones** of the wrist, and the five **metacarpals** and fourteen **phalanges** of the hand. These

Process

Years ago, I was overlooked for a role in a dance because I did not have classically curved arms. This decision motivated my visit to a bodyworker to begin addressing the tension in my shoulder girdle. After a one-hour session of gentle manipulation and repatterning, I left relaxed and seemingly unaffected. The next day my arm was very sore. As that subsided, my neck went into a spasm which lasted a week. When that passed, my left arm began to ache. I began to realize that the work was travelling through my body. For several months, my left leg would go numb whenever I was tense. Sometimes when I was performing, I wouldn't be able to feel my foot touch the floor. Eventually, this too left and the process was complete; it had moved through the skeleton and reached the ground.

Sculpture: Gordon Thorne
"Air"

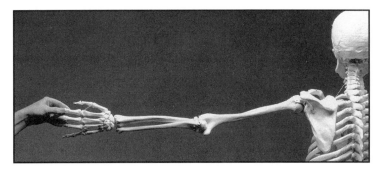

Shoulder girdle: posterior view of left arm

peripheral bones provide a levering system into the center of the body for force and for shock absorption. Two portions of the scapula extend from the back to the front surface of the body: the acromion, and the bony coracoid process which protrudes forward, under the acromion, for muscle attachments. If all of the bones involved in the shoulder girdle were extended laterally, we wouldn't be able to walk through a door! The main problem with the front to back wrapping is that we often forget to involve the scapula in the action, connecting the humerus instead directly to the clavicle and slightly dislocating the shoulder, causing various discomforts. The humerus should articulate evenly with the entire surface of the cup of the glenoid fossa. *The only joint connecting the shoulder girdle to the axial skeleton is between the clavicle and the manubrium (see page 40).* This is a fluid-filled, synovial joint, and allows free movement of the shoulder girdle separate from the ribs. The shoulder girdle along with the

Scapula: anterior view of right shoulder joint and ligaments, ribs removed

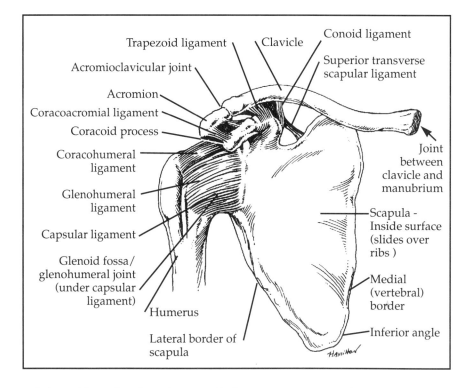

pelvis and legs developed as weight bearing structures after our evolution to land and constitute the appendicular skeleton.

If we trace the levering from the fingers through the shoulder girdle and the skeleton when catching a ball, we see that the impact goes into the wrists, around and through the radius and ulna to the elbow, up the humerus, into the glenoid fossa (not the clavicle) which takes the force to the back of the body into the scapula, down to the inferior angle and around the medial border, the spine and the acromion, then into the clavicle on the front surface (through the acromioclavicular joint), and into the axial skeleton at the manubrium. It then travels down the sternum, out and around each of the attaching ribs, and into the transverse processes, bodies and discs of the vertebrae. There it travels down the front of the spine and spreads into the pelvis, which generally provides equal and opposite rotational force if the ball is caught in a spiral. The force travels down through the legs to the feet and into the earth. Holding in any of the joints causes stress. The ideal is even and equal distribution of weight or force at each joint. For ex-

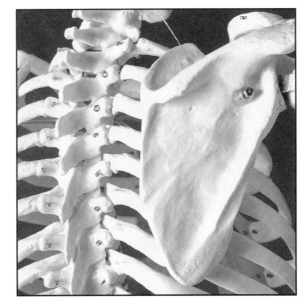

Scapula: posterior view
Acromion process
Spine of scapula
Supraspinous fossa
Infraspinous fossa
Medial border
Inferior angle
Lateral border
Glenoid fossa
Coracoid process

ample, shoulder injuries for racquetball or tennis players can be caused by locking the elbow before impact, thus sending all the force to the next available joint, the shoulder. Stability of muscles at a joint is necessary, but skipping a joint can cause stress. Efficient use of the arm involves clear levering from the periphery to the back of the body (scapula) to the front of the body (clavicle, manubrium, sternum) to the center of the body (through the ribs to the bodies of the vertebrae) to the earth.

The nerves for the arms and hands pass between the cervical vertebrae in their journey from the spinal cord to the periphery. Tension in the neck can cause pain or numbness in the arm and hand; balance of the neck on the spine can reduce arm stress. When working with the shoulder girdle, check postural alignment of the three body weights over the base of support. Integrated alignment of the axial skeleton: the skull, spine, ribs and hyoid bone, is the foundation for efficiency of movement in the appendicular skeleton: the shoulder and pelvic girdles.

Movement can draw us towards center, towards our axial skeleton; or into space, through our appendicular skeleton. Proximal initiation is movement initiated by bones or joints closest to the center of the body. It will bring your movement towards your center. Distal initiation is movement initiated by the bones furthest from center and will take you into space. For example, if you initiate a movement from your shoulder, your movement stays close to the body; if you initiate from your hand, the movement takes you into space. Try crawling on your hands and knees: reach with your shoulder as you crawl; reach with your hand. Both are useful, but they have different results. ❖

Shoulder circles
20 minutes

Lying on one side, legs bent slightly for support, arm under body resting comfortably supporting the head like a pillow or relaxed:

O Leading with the fingers, draw the top arm up towards the ceiling, perpendicular to the body. Close eyes to increase proprioceptive awareness; feel the humerus drop into the scapula for a base of support for the arm.

O Begin drawing tiny circles on the ceiling with fingers.

O Gradually allow the circles to enlarge, moving very slowly. Imagine using little or no muscle. This will stimulate work with the deep muscles (intrinsic) rather than the superficial muscles (extrinsic).

O When the circles become comfortably large, slowly reverse the direction.

O Gradually draw smaller and smaller circles on the ceiling with your finger tips.

O Inscribe the smallest circle possible on the ceiling, feeling the vertical axis.

O When finished, relax fingers towards gravity; relax the wrist; relax the elbow; carefully relax the shoulder and place the palm of the hand on the floor in front of your ribs.

O Press finger tips and palm into floor and connect into the scapula to push yourself up to seated position.

O Stand, feel both arms. Do other side.

Tracing the shoulder girdle
10 minutes

With a partner or alone:

Lying in constructive rest: Place your fingers on the joint between the clavicle and the manubrium and feel its shape.

O Walk your fingers out along the clavicle, fingers on top of the bone and thumb below it to feel its width. You will find the acromioclavicular joint approximately where the strap of a shirt would cross your shoulder.

O Continue on around the acromion (like an epaulette on a military uniform) as far as you can on the back surface of the body where it becomes the spine of the scapula.

O Return along the bone to the front of the acromion. Like jumping off a cliff, keep your fingers on the acromion and let your thumb jump off and find the head of the humerus (the top of the arm). It is often protruding quite close to and in front of the clavicle. See if you can feel any space between the two bones and trace the top of the humerus.

*Partner work: stimulating the shoulder girdle**
15 minutes each arm

○ Continue on with the previous work by lifting your partner's elbow with one hand while keeping the other on the head of the humerus and the acromion.

○ Supporting a slightly bent elbow, slowly begin circling the head of the humerus in the glenoid fossa -- the shoulder joint. Feel the connection between the elbow and the head of the humerus as a line leading into the socket of the scapula. This may take some time. Continue to move slowly. If you feel muscle resistance, move in a different direction. It is not useful to push through muscle tension, it will only generate more tension. The goal is to move the bone underneath the muscles, allowing them to relax so that you can stimulate the joint surface evenly. You may also feel small jerks as nerves release; this is actually a good sign of relaxation.

○ When you feel ready, bring the elbow up towards the head. You may need to move so that you are sitting on a diagonal facing the belly. Gently lever (push, compress) the humerus into the joint and watch the movement travel through the bones all the way into the center of the body. This is joint stimulation through compression.

○ Then reverse your work and elongate the arm slightly away from the socket for stimulation through elongation. Your partner can give you feedback; there should be no pain.

○ Return the arm to the side of the body; complete your work with a few slow circles and place the arm carefully on the ribs. Move to the other side of the body and draw the hand across the torso, asking your partner to roll on her/his side as you do so.

○ Press the fingers and palm into the floor and ask your partner to press into the hand to come to seated. Take time to reorient to eyes open; stand and feel both arms. Repeat on the other side.

○ When you finish, have your partner walk briskly, easily swinging both arms in a natural walking pattern. Move faster than you can think, approximately two steps per second, so you are integrating the work into normal movement patterns.

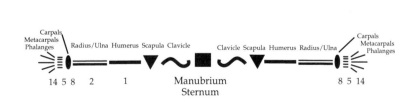

* This exercise is based on Touch and Repatterning techniques of Bonnie Bainbridge Cohen and The School for Body-Mind Centering.

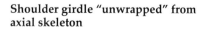
Notes:

If you repeat this excercise, you can also locate the coracoid process, the protuberance of the scapula which reaches forward to the front surface of the body for muscle attachments. It is a small knob located below the clavicle in the indentation where a thin strap would cross your shoulder.

If working with someone who has an injury or sensitivity in the shoulder joint, begin with the well arm. You become clearer in your touch, and the nervous system will transfer the information to the other side, facilitating your work.

Always do both sides. Find a way to bring your partner into normal activity to stimulate the protective neuromuscular responses necessary in daily life. Suggest to your partner that he/she avoid heavy lifting or strenuous exercise directly after this work; the joints are open and injury to ligaments can occur with outside weight or force.

Shoulder girdle "unwrapped" from axial skeleton

Moving the hand from the scapula

If you are with a partner, place your right hand on their right scapula. The little finger relates to the inferior angle. The fourth finger to the lateral border. The third finger to the joint/glenoid fossa. The pointer finger to the spine of the scapula. The thumb to the coracoid process.

○ Have your partner swing their hand and feel, encourage, the movement of the scapula. Go through each finger and its corresponding area of the scapula with one full pendular swing for each finger. You may find that the swing initiated from the little finger travels on a diagonal path, whereas the thumb initiates an up-and-over curve. Feel the connection between the support of the scapula on your back surface with the movement of the hand in space.

○ Release your hand from your partner's back and try this exercise yourself: leading with each finger, swinging the hand, and feeling the movement of the scapula.

○ Shake hands with someone, reach with the scapula. Reach without using the scapula to feel the contrast of support.

○ Hug someone or yourself; feel the reach with the whole shoulder girdle. Hug holding the joints to feel the contrast.

○ Push against a wall or a partner with both hands; feel the levering travel all the way down to your feet.

○ In constructive rest, visualize that your arms are hoses and water is flowing from your chest through your arms and draining out your fingers into the earth. Image that the water flows through the neck and into the skull, out the top of the head and down the torso and out the pelvis and feet.

FOREARM, WRIST, AND HAND

The study of the forearm and hand returns us to our evolutionary forebearers. In the transition from sea to land, fins became weight bearing appendages (as in the salamander) and gradually folded towards the midline, lifting the belly off the earth to reduce friction in locomotion. Rotation of body parts prepared the body for efficient forward propulsion, lining up head, hands, feet in the direction of motion. Thus, in many of our four-footed ancestors, the two bones of the forearm crossed over, flipping the palms down and the fingers forward. (Our familiar friends, the cat and dog, have paws that push effectively against the earth to propel the body forward. A horse, even higher up on its digits, walks on its finger tips and toes, adding propulsion and speed but losing agility and spring. The gorilla kept the backs of the hands down and walks on its knuckles.) This rotation, as we have already noted, allowed the body to lift off of the ground: the vulnerable belly surface, once safely in contact with the earth, became exposed. When we rotate our palms down for crawling, we can observe the crossing of the forearm bones, and the past distinction between back and belly surface.

Our hands evolved from paws, changing their function from weight-bearing and propulsion to articulation and manipulation. As we began

Control

For ten years I have worked in a process called Authentic Movement introduced to me by Movement Therapist, Janet Adler. In this work, there is a witness and a mover. The mover closes her/his eyes and waits for movement impulses in the body – the process of being moved. The witness holds the consciousness and observes. After a period of time, the witness calls the movement session to a close, and there is verbal exchange about what occurred.*

❖

For weeks, I would lie on the studio floor and find myself reaching with my right arm, wiping it horizontally in front of my body. I recognized this movement as a gesture I had incorporated in many of my dances, combining it with turns or leaps, or bringing it to the floor. But I was surprised it was so determined to appear in my Authentic Movement sessions. (Later Janet said, there is no need to "remember" movement from one session to the next. If a gesture wants to be recognized, it will return again and again until it is brought to consciousness.) One day when this movement appeared, I got the image of myself as a little girl in our farmhouse in Illinois. We had a long counter top in our kitchen which collected objects, and it was always a mess in the usual family way. Whenever my parents would argue, I would clear the counter. It was my way of controlling, quieting, bringing order to what I perceived as chaos. The movement pattern in my body was sourced in this childhood activity. And I was still using it in the same way, to bring order and calm (and the safety of the Illinois landscape) to the dynamics of performance.

Writing Table: Kristina Madsen

❖

The work in Authentic Movement is valuable in many ways. In this situation, it gave me choice about the use of a movement in choreography. The hand-wiping was not a particularly interesting gesture to watch, and I probably didn't need it in all of my dances. It was compelling to both the audience and to myself primarily because of my investment in its content. As I brought awareness to this language of the body, I began to listen for other patterns based in my personal story.

hunkering (squatting) in tree branches, our hands were free for picking berries and nuts, grooming, and air-born locomotion. The nervous system refined so that both face and hands became highly expressive tools for survival and communication. The thumb in this transition, developed an oppositional pattern with the fingers for grasping and tree-swinging. In the human baby, we can reobserve this progression in the change from the newborn's generalized hand to the strengthening of thumb opposite fingers for grasping.

The forearm is composed of two bones, the radius and the ulna. The radius, on the thumb-side of the arm, is large at the wrist and becomes smaller at the elbow where it creates a pivot joint with the humerus to allow the crossing of the forearm. The ulna, on the little finger side of the arm, serves as a small brace for lateral support at the wrist, but becomes the primary articulating bone at the elbow. It works as a hinge joint with the humerus, and can be the site of pain for individuals with "bony elbows." It is important to understand that the hinge joint has curved surfaces, like a door hinge, rather than flat surfaces, like a folded piece of paper.

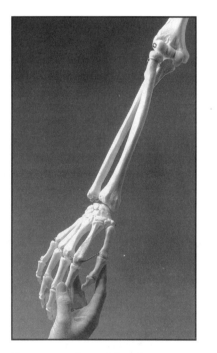

Forearm, wrist and hand: anterior view of right elbow; posterior view of wrist and hand

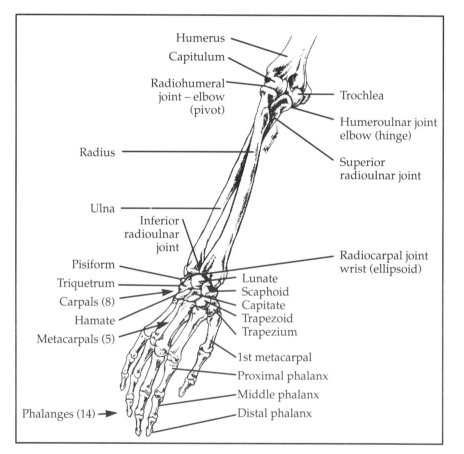

Weight or pressure from the hand is levered through the wrist to the radius, where it travels around through its skeletal connection to the ulna and up through the elbow to the humerus. Thus, the radius is the primary bone at the wrist and the ulna the primary bone at the elbow. An interosseous membrane weaves the two bones together with strings of tough connective tissue like the webbing of snow shoes. This structure gives the forearm lightness (imagine if it was one solid bone!) and mobility, and provides attachment sites for the many muscles of the hands.

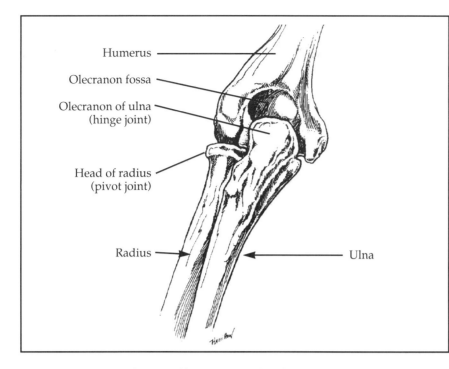

- Humerus
- Olecranon fossa
- Olecranon of ulna (hinge joint)
- Head of radius (pivot joint)
- Radius
- Ulna

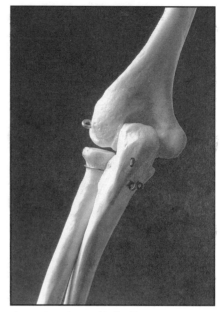

Posterior view of left elbow

There are eight small, irregularly shaped bones of the wrist collectively called the carpal bones. Their surfaces glide slightly over each other, allowing three-dimensional movement. As you circle your wrist, you can feel the tiny jerking movements from the articulating bones; this structure provides mobility, while maintaining stability with minimal musculature. The carpals also collectively create a curved surface which articulates with the radius. Thus, the wrist is considered to be an ellipsoid joint between the carpals and the radius, and a collection of gliding joints between the carpals. There is also movement between the carpals and the five long bones of the hands, the metacarpals. ❖

❖

We were sitting around our campfire near the Masai Mara game reserve in Kenya with our guide Patrick Pape. He is a man of patience and integrity who embodies the adventuring spirit. He told a story about his assistant on tented safaris. This individual prided himself on elegance and order in camp, but was prone to accidents. He had lost a finger while loading gear onto a large truck, and was most distressed that his hand would be unsightly when serving meals. Patrick bought a pair of white gloves and stuffed one finger with cotton. Soon all of the meals were served by a staff wearing white gloves.

* For further information about Authentic Movement, see Janet Adler's article "Who is the Witness?" *Contact Quarterly*, Vol. XII, No. 1. 1987.

Rotating the forearm; tracing the bones of forearm and wrist
10 minutes

In seated position: Place your hands, palms up, on your legs in front of you. Stabilizing the elbow, rotate the forearm so that the palm goes down to face the leg without moving your elbow. Note that the pivot for the crossing of the forearms occurs at the elbow.

○ Hold the radius and ulna with one hand; gently circle the eight carpals as a unit in relation to the forearm. Hold the carpals; circle feeling the subtle movement between the eight bones. Still holding the carpals, circle the hand in relation to the wrist. Note that you have all three levels of articulation for shock absorption and refined movement at the wrist.

○ Beginning at the thumb side of the wrist, feel the shape of the radius and trace the bone up to its pivot connection at the elbow.

○ Return to the little finger side of the wrist. Trace the ulna up to its hinge connection with the humerus at the elbow. Then, using thumb opposite fingers of the touching hand, trace the space between the two bones from the wrist to the elbow. This action stimulates the interosseous membrane and the fluids.

HAND

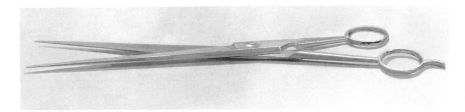

Painting: Jim Butler
Untitled

The hand is a highly articulate structure, carrying out the commands of the brain and receiving information from the environment. Along with the eight carpal bones of the wrist, the hand is comprised of five long metacarpals, and fourteen small phalanges – two in the thumb and three in each of the other four digits. Each bone of the fingers articulates as a hinge; (like a door, moving in two planes of action) the base of the thumb is considered a saddle joint because it has a curved surface allowing circular movement and oppositional crossing with the fingers.

The refined skills of pianists, surgeons, Balinese dancers, and rock climbers attest to the potential for highly specific movement in our periphery. This is due to the quantity of small muscles in the hand (the foot and the face also), and the high ratio of nerve endings to muscle fibers making subtle movement changes possible. Like the eye, the hand is a highly sensitive communicator of internal processes and receptor of the outer environment. ❖

Tapping

A friend of mine used to tap her pencil constantly. At meetings it was the undercurrent of all activity. We began working in movement. She told me she had been labeled hyperactive as a child, and had even worn a harness so her parents could keep her in sight. As we moved, eyes closed, a multitude of movement came out – endless sounds and stories and memories and gestures normally held inside. Now, whenever I see someone tapping a pencil, I smile. What would happen if they closed their eyes and moved.

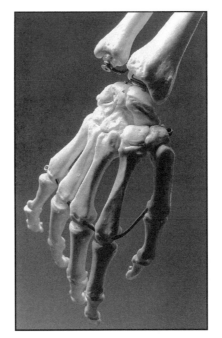

The Hand:

8 Carpals

5 Metacarpals

14 Phalanges

Hand massage
10 minutes

In seated position (or in constructive rest with a partner), rest the back of the hand on your thigh; make sure the elbow is slightly bent and supported.

○ Begin circling the little finger, slowly stimulating the joint between the bottom phalanx and the metacarpal. Continue with each of the fingers and the thumb.

○ Begin pincering (squeezing) down the space between the first and second metacarpals using your thumb on the palm side of the hand and the pointer finger on the back surface.

○ Continue tracing the space between each of the metacarpals, from fingers to wrist, encouraging even and equal space between each bone.

○ When you reach the little finger side – the fifth metacarpal – begin gently massaging the eight carpal bones, feeling for their shape and their mobility. Continue into the space between the radius and the ulna.

○ To finish, generally massage the palm of the hand, moving the fluids and integrating the sensations. Repeat with other hand.

Shoulder girdle integration: heart circle
5 minutes

In seated position, press the palms of your hands together in front of your sternum, elbows lifted out to the side.

○ Let each finger tip press into its pair (right thumb to left thumb) as well as having the palms touching; fingers can point up or forward as is comfortable to you. Feel the pressure go evenly through each set of metacarpals, carpals, and into the radius and then the ulna, into the humerus and the glenoid fossa.

○ Check at this point to make sure the force travels into the scapula in back, then forward around the spine and acromion of the scapula to meet the clavicle in front. Here it connects to the manubrium, and travels into the sternum, around the ribs to the bodies of the vertebrae. Now feel the warmth of the hands connecting to the warmth of your heart and lungs, periphery to center.

○ Breathe deeply. If seated you can continue to transfer the force down the bodies of the vertebrae to the pelvis and into the floor. *

○ You may do the same with a push up: lever from the hands into the center of the body, traveling through the skeleton and including the heart, lungs and breath. Be sure to include the scapulae; even though you will need to stabilize these bones wide on the back for support, the force still travels from each humerus into the socket, around the scapula and into the clavicle, sternum and ribs to the spine.

* This can later be done standing, carrying the force down through the legs to the floor.

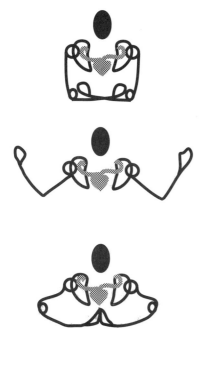

Heart circle: yes, no and connection

Specifically for dancers: carriage of the arms

Standing with arms open to the side in second position:

❍ Feel the humerus in the socket. Rotate the bones so the elbow and wrist face back, the front surface of the arm and the palm face forward. Image the energy traveling through all the joints out the middle finger. (There is no crossing of the radius and ulna, the bones are parallel to the earth.)

❍ Energize and maintain this relationship as you bring the arms to fifth position en haute, overhead, palms facing each other. Rather than lifting your body up as you raise the arms, the weight travels down, draining into the scapula, and continuing around the shoulder girdle, to the bodies of the vertebrae and to the ground. This use of the scapula as a counterweight creates the illusion that the arms are floating. If the whole body tries to float, there is no equal and opposite force for stability.

❍ As you begin to lift the arms, allow the humerus to rotate gradually in the glenoid fossa, and feel the scapulae spread wide on your back to make feet, or platforms, to support the arms overhead. This greatly reduces muscle use around shoulders and neck, and creates a circle from the center of your thorax, your heart, out and around the arms, which connects through the space between your finger tips. Whether the circle is directly overhead, or slightly forward depends on the structure of your shoulder girdle,* and your personal aesthetic. Remember that the entire shoulder girdle pivots from the joint between the clavicle and the manubrium and wraps around the body, free of bone restrictions.

❍ Reverse the action, opening the arms into second by sliding the scapulae on the back, rotating each humerus in the socket, and extending through the elbows, wrists and palms out the fingers. The movement involves continual rotation for a smooth transition. In second, the elbow and back of the hand face back; the palm forward with a slight, continuous curve through all of the joints as if you are hugging a giant balloon. Your finger tips are visible in your peripheral vision as you look straight forward. Energize the surface of the arm. Feel the wrapping of the shoulder girdle as you connect the humerus back into the scapula, then around to the front to the sternum, then back through the ribs and front, down the bodies of the vertebrae, to the pelvis, the legs, the earth.

❍ As you lower your arms from second to low first, you again move the scapula and rotate the humerus in its socket. You may choose to add a rotation through the wrist and hand for gestural affect, crossing the radius and ulna, and extending out the finger tips for a graceful, flowing action as the hands lower to their position slightly in front of the thighs, palms up.

❍ We finish with the circle of energy from thorax out through hands, framing the torso in a low position.

* The size and positioning of the scapulae vary from person to person.

TO DO

Note:

Porte de bras, or carriage of the arms, for ballet or modern dance requires an efficient levering system through the shoulder girdle. It also involves subtle rotations of the humerus in the glenoid fossa, which are simultaneously carried on through the radius and ulna and hand for gestural refinement.

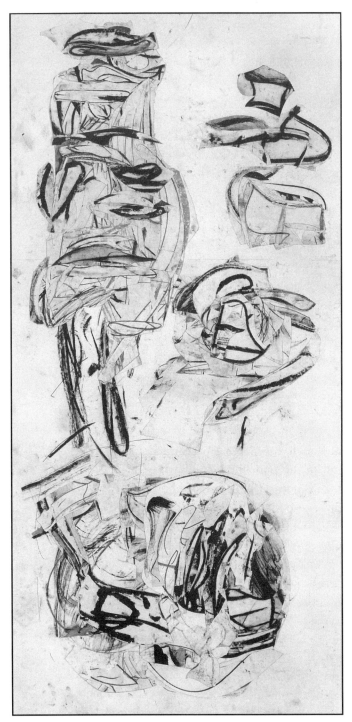

Drawing: Michael Singer
"Ritual Series"

TOUCH

Touch is our connection to the world. We can see, hear, think about something, but it is through touch that it becomes part of our experience. Touch informs our perceptions, stimulates our reactions. Its effects vary with our intention. We can place our hand on someone's shoulder and feel the bone, muscles, fluids beneath the skin. A hand on a shoulder can also feel nothing. We can place a hand on a shoulder to stimulate a sensual response. A hand can also be put on a shoulder because someone needs something. The same activity of touching communicates very different intentions and evokes different responses. How we touch ourselves informs our body attitude. Do we brush our hand on our face and feel detached, or feel anger? We can touch a body part with judgment or with pleasure. Observe the qualities of how you treat your own body; imagine how you would like to be touched.

There are many ways of touching. In bodywork, touch is a dialogue. Place your hands on a body part, and "wait." Bring your attention to the area and receive information through your hand, like a sensitive microphone. You are creating a dialogue between your hand and the area being touched. In this "empty" time of not doing, you are actually bringing awareness and initiating the dialogue which can affect change.

The more intentional your touch, the clearer you can feel in the hand doing the touching as well as in the part being touched. Touch creates sensation. Random proprioceptive input is like static in a recording, it distracts rather than contributes to your ability to hear. For example if you are touching a partner to feel the bone in their forearm and you are pressing very hard, you may have so much stimulation in your own body that you can't feel what is happening in the other person. This is similar to having a conversation where you do all the talking; all you hear is yourself. Sometimes in hands-on work our desire to be helpful, to *do something* for someone, interferes with our actual ability to hear what needs to happen.

Attention and intention are key to hands-on work. Our choice of touch is informed by our intention. If we are working with bone, we might begin with a firm direct touch which penetrates the layers of soft tissue; the intention of feeling for bone guides our touch. If we want to relax muscles, we might use a muscular movement of the hand to stimulate the fluids and connect us to muscle. We can often intuit the most efficient touch as we work more effectively than deciding ahead of time. We are *per*forming rather than *pre*forming touch. And, in a way, we are training our intuitions

Staying Present

When I showed a fellow bodyworker the first draft of my writings, we went page by page through the manuscript. As we reached the section about touch, her face changed. I had described "Active" and "Passive" touch as ways of working. She said: "There are very few situations when I would encourage someone to be truly passive when they are being touched. When someone is relaxed, they can still be present. When I am working with someone, I rely on being met; I experience touch as an exchange. Even when someone is giving me a massage, my intention is to meet the touch, rather than having something done to me in a passive state. It is important training for life."

Body Picture

"I hate my feet," a student said adamantly. She was extremely ticklish and hated being touched or touching her feet or ankles. She told me that she had worn braces on her feet as a young child, consisting of a metal bar between stiff shoes, to correct a problem with her ankles. She described with residual frustration how the device had made it extremely difficult for her to crawl, to stand, or get to the bathroom in time to pee. She had been labeled a "bed-wetter," and it was all because of her feet. She was now a skilled athlete. Her favorite sport was swimming, which focused on the upper body. We used sensory stimulation to "wake up" her lower legs. As well as the soothing touch of massage, we lightly scratched her legs and feet to arouse sensation by slight irritation. After our work, she stood up and could feel her heels and toes on the ground. On the last day of class, she took off her shoes and socks and said, "I think I like my feet."

❖

Another student in the same class had knee surgery two years before, and was still confused and irritated by knee pain. When she performed as a dancer, she took all the stress in her knees, with little awareness of her lower leg. She too had worn the foot brace as a child. As we worked, she realized she had minimal sensation from her knee down, and that her image of her body stopped at her knee. Because her feet had been immobilized, she initiated movement with her knees.

❖

A young athlete had been in two foot casts as a baby to correct a tendency towards being "pigeon-toed." His pain was now in his lower back. Because he couldn't use his feet, all of his movement had localized in his spine. Now when he walks, he still jams into his lower back, rather than using the leg and feet muscles to cushion and sequence the movement.

❖

I became aware that six of eighteen students in this particular class had worn corrective bars or feet casts when they were babies or as young children. It was a common medical practice at the time. All benefited by "normal" looking feet, but suffered from injuries in related joints of the spine, hip, knee, or ankle and had a confused proprioceptive image of their lower legs. All were frustrated with the lack of mobility or stability (balance) in their body. All were ready for change.

as we work. Pain is rarely useful; it causes muscles to contract and blocks responsiveness of tissue. Pain can, however, draw attention to a place overlooked with subtler messages and encourage awareness of areas in need of health. Judgment, self-criticism interrupt the capacity to feel the body through the body. In general, what would feel good to you will feel good to someone else. Your attention remains on the place you are touching and its connection in your own body; your intention is to listen and respond to what you feel happening beneath your hands.

TYPES OF TOUCH

Each person develops individual touch techniques. As we have said before, the intention of touch shapes its effect. There are, however, general types of touch which can be useful resources. Try them on your thigh or forearm as you read. Touch through cloth or on the skin; the cloth adds additional sensory input and friction, sometimes useful in work.

Pressure with palm or flat surfaces of fingers: This direct, firm touch provides a comforting, secure pressure which allows deep work without stimulating reflexive contraction in muscles from pain or "ticklish" sensation. (Tickling is random stimulation of nerve endings, registering confusion in the cerebellum.) Firm, direct pressure over large areas is reassuring and allows clear patterns to be perceived, received.

Pressure with finger tips, thumb: This pointed pressure can go deeper through tissue layers because of minimal surface contact and direct leverage through the bone structures. It is good for deep release of specific areas, but can cause reflex contraction from pain, tickling. It is good to follow deep pressure with generalized rubbing or scratching to integrate proprioceptive awareness with a larger area.

Brushing, scratching: This technique stimulates surface proprioceptors. It can be useful to bring generalized awareness and sensation to an entire area. It can be done effectively as a stroking, pleasurable sensation, or in a scratchy, slightly irritating way. Both are useful. Sometimes the gentle irritation can be more effective to stimulate an area which needs to be "woken up" to the body picture (an area numb from injury, fear, lack of use). Variations in type of texture of brush used can be helpful (toothbrush, soft hair brush) as well as the hand.

Diagonal pressure across fibers: This sliding of tissue over tissue separates connective tissue from muscle fibers, and fibers from each other. Keep sensations at a comfortable level. Used by experienced Rolfers, this technique can stimulate mobility at every level of body structure.*

Pulsing, pumping: This technique primarily affects fluids in the body which collect due to stress, injury, or inactivity. Pulsing or pumping the hands along a body part can stimulate the fluids in the soft tissue, like a gentle squeeze of a sponge or toothpaste tube. This is especially useful from the thigh to the foot, or shoulder to hand to bring fluids to the periphery; or the reverse, to stimulate the flow towards center if fluids are pooling in the periphery.

Bone manipulation: Articulating the surfaces of bone ends carefully in their joint through compression elongation, rotation, flexion and extension, circumduction as is appropriate for the specific joint stimulates proprioception in the joint capsule and can release muscle tension. This is a very important technique for neurological repatterning and should be used with care and specific instruction.

Tendon/ligament support: Finding the articulating surfaces of bone and the origin and insertion of ligaments can bring some reorganization to ligaments and tendons which have been damaged or misaligned in injury. It is important to clearly visualize the direction of fibers, place fingers on the origin and insertion of the ligaments, and wait. Your fingers provide support for the joint similar to the ligaments and stimulate their receptors. You will need a diagram of the ligaments. Try this at the TMJ or at the knee.

COMPLETING YOUR WORK

Balance the body; do equal work on paired body parts. Take time to bring vision back into awareness. Transition carefully from inner experience with eyes closed, to reintegration into an outer environment with eyes open. Stand up sequentially, feeling the shifts in postural alignment.

Use speaking, drawing, walking, playing (tossing a ball) to bring the senses back into generalized focus. Integration is part of the process. Never leave your partner feeling disoriented. Walking at a pace of two steps per second (faster than you can think) shifts awareness into activity.

Allow time before doing any vigorous activities: lifting weights, sports, performing, etc. Bodywork can open joints beyond their usual range. Establish an ongoing dialogue of trust between you and your body. It will open if it trusts you will be responsible for and responsive to the changes that occur. Otherwise, muscles will have to regrip to protect you from yourself!

INJURY

Injury or illness can be seen as the body's way of calling attention to a particular area. It often reflects misalignment or stress which has been occurring for a long time; when an impact occurs, it settles in our weakest area. This "weak" area is often, in actuality, a place of strength that we use to exhaustion, injure, and then label as "weak." For example, a runner's knee that at first enables the pleasure of running, then becomes stressed by long distance repetition of that same pleasurable action, and finally becomes a "weak" knee, susceptible to injury on impact, in a car accident, or in a fluke, awkward moment of tripping. Rehabilitation involves recognizing the relationship between our weaknesses and our strengths, and attending to the underlying patterns as well as healing a particular area. A problem can be an invitation to learning. ❖

Hands On

I attended a lecture given by Professor John Truxal of the Department of Technology and Society at State University of New York called "Demystifying Technology." He began by passing around a pacemaker for each of the audience members to hold. "There is nothing very complex about technology," he said. "We made it." And he laughed. "Now, the human body – that's complex."

---

* See Don Johnson's *The Protean Body, A Rolfer's View of Human Flexibility.*

Touch: sensitizing the hands
5 minutes

Rub your palms together to generate warmth. Then press your fingertips together, gradually connecting through the bones until the palms are touching. Feel the gentle pressure or "levering" through the bones from your finger tips through the hands, radius and ulna, humerus, scapula, clavicle, manubrium and sternum, ribs, vertebrae and into the soft tissues/organs of the lungs and heart. This "**heart circle**" from the palms to the center of the body brings blood flow and warmth to the periphery.

O Gradually separate the hands and stay aware of the heat and energy in the space between them. Move them as far apart as you can, maintaining the sensitivity. Recognize the proprioceptive and cellular awareness. Bring your concentration to your hands, to the space between the hands, and to the connection of hand to heart (blood flow). Experiment with degrees of sensation: energy and heat without actual touch, barely touching, deep pressure.

O In seated position, place your hand on your thigh. Feel the touch of cloth against your palm. Bring your awareness to the skin surface and warmth below the cloth. Shift your focus to the denser tissue layers of muscle and fascia. Reach your mind's eye to bone – the densest tissue. Apply deep pressure and feel through the layers. Return to light touch and feel for the various densities of tissue. You may also notice the pulse of blood, electrical currents of the nervous system, muscle twitches or responses. Allow yourself to experience whatever is present in the moment through your senses. Eventually you will be able to differentiate various tissues through touch.

Touch: working with a partner

Sitting in a comfortable position close to your partner: Find your plumb line and establish a secure base of support. Connect your weight to the ground through your pelvis, so electrical impulses and body energies can pass through your structure to the earth.

O Breathe evenly and fully so that your body is responsive. As you work, follow the experience in your own body. If you are touching someone's spine, feel your own spine. Monitor the amount of force and stimulation needed as you work; remember that the more "active" your touch the more you feel yourself.

O When you finish, bring both you and your partner back to a neutral state; separate your energy connection to your partner by touching the floor or washing your hands. Use the "progression to standing" to come to vertical; have your partner walk at a quick pace; move or talk together to integrate the experience before returning to daily activities.

PELVIS

The pelvis is a bowl composed of circles, holes and arches. It holds the digestive and reproductive organs and creates a passageway for the birth canal and for elimination of digestive wastes. It also provides paths for the nerves and blood vessels traveling to the lower limbs, and serves as a site of attachment for the many muscles integrating the axial and the lower appendicular skeletons. The pelvic bowl is free-swinging around the two femur heads of the legs, and connects at the back to the sacrum and spine. It is one of the three primary body weights, aligning with the skull and thorax around a vertical axis for efficient posture. It is a highly mobile structure, constantly responding to activity below and above.

The pelvic girdle is formed of two matched halves, connected posteriorly to the sacrum and anteriorly by the pubic bones. Each half consists of three bones, an ilium, ischium and pubic bone, which fuse together to provide stability at the hip socket, the acetabulum, early in a child's development. The two ilium bones fan into wide arches from the sacrum, forming the bowl in the back; the ischial tuberosities are the bottom "feet" of the pelvis and transfer weight to the chair or floor when sitting; and the pubic bones meet to create a forward thrust, like the prow of a ship, for attachment of the abdominal muscles and fascia. All of the bones follow curved pathways and transfer weight around and through their structures to other bones. There are six moveable joints in the pelvis: the two sacroiliac joints between the single sacrum and the two ilia, the two hip joints between the two acetabula and their femurs (thigh bones), and the pubic symphysis between the two pubic bones and their disc.

Forgetting

I had a student very interested in body work in a college course. When we got to the exam on the skeleton, she missed every question about the pelvis. When I asked her about this, she said she couldn't remember the names of the bones, couldn't draw the shapes. Later, she told me she had been sexually abused in high school; that she was afraid to articulate her feelings. I asked her if that was why she couldn't remember the names of the bones of the pelvis. She began to realize the connection. When it was time to give final grades, I felt the dilemma of testing. Often what a student can't learn, gets wrong, is the key to their learning.

*Theatre Director Joseph Chaiken was teaching a workshop. One of the performers from **For Colored Girls Who Considered Suicide When The Rainbow Wasn't Enuf** was talking about her experience as an actress. "When I am learning a script with Joe," she said, "we always circle the lines and words that I can't remember. Then we go to work on these areas to unfold my own story within the character. The blocks are the places where the personal connections will come."*

**Drawing: Harriet Brickman
"Beached Forms Series"**

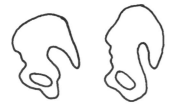

The Pelvis: female (left) and male (right) shapes and photo below, left lateral views

Together these joints absorb the impact of walking to protect the vital contents of the pelvic bowl. Each acetabulum is composed equally of one third ilium, one third ischium, and one third pubic bone, like a concave pie divided in three pieces. This allows equal force from all three directions of the bowl to pass into the hip socket. The pelvic bowl includes the sacrum, the keystone and the central link between the axial and the lower appendicular skeletons. The many holes in the pelvis allow its lightness, and the curves and moveable joints transfer stresses throughout the structure to neutralize impact. Once again, the mobility of the structure is its stability.

The pelvic floor is a horizontal diaphragm formed of ligaments and muscle tissue. It supports the organs and forms the mobile "bottom" of the pelvis, interweaving the pubis with the ischial tuberosities with the sacrum and coccyx. Maintenance of the tone of the pelvic floor is important for organ support and childbirth. Other horizontal supports through the body which parallel the pelvic diaphragm are the cranial, vocal, and thoracic diaphragms. ❖

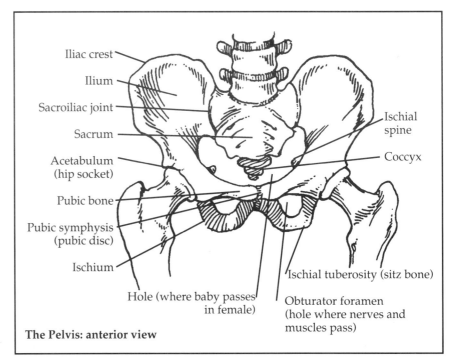

Iliac crest
Ilium
Sacroiliac joint
Sacrum
Acetabulum (hip socket)
Pubic bone
Pubic symphysis (pubic disc)
Ischium
Hole (where baby passes in female)
Ischial spine
Coccyx
Ischial tuberosity (sitz bone)
Obturator foramen (hole where nerves and muscles pass)

The Pelvis: anterior view

Pelvis
30 minutes

Lying in constructive rest, hands on your belly:

❍ Trace the bones of the pelvis: Starting at the iliac crest below the ribs, walk your fingers forward until you come to the thrust of the pubic bone and the pubic symphysis.

❍ Roll on your side: Starting again at the iliac crest, trace the bone back to the sacrum; feel the sacroiliac joint. Massage through the soft tissues of the gluteal muscles to the acetabulum (hip joint), and locate the neck of the femur and the greater trochanter. Continue down the back of the pelvis and locate the ischial tuberosities (the "sitz" bones), the bones you sit on in a chair. Flex the top leg if necessary, and trace from the sitz bones to the pubic bone between the legs. This is called the ramus of the pubic bone. Roll to the other side and repeat.

❍ Seated: Rock on your sitz bones to feel their shape against a chair or the floor. Image them as the feet of the pelvis. Roll to the side and feel the space between the sitz bones and the greater trochanter of the leg.

❍ Return to center and again rock forward and backwards on the sitz bones. Initiate the rocking from the iliac crest; initiate the rocking from the pubic bones; initiate the rocking from the ischial tuberosities (the sitz bones), initiate the rocking from the pelvic floor. Circle the pelvis, imaging swishing water around the inside of the pelvic bowl; reverse directions.

❍ Seated: Walk on your ischial tuberosities, moving forward in space. Keep the legs relaxed. Reverse directions. Do the same thing, thinking of separating the pelvic halves. Walk forward initiating from the iliac crests. Walk backwards initiating from the ischial tuberosities. Walk forwards initiating from the pubic bones. Walk backwards thinking of the hole.

❍ Lying on the floor: Roll, initiating movement from the pelvis. Reverse. Roll, initiating movement from the organs, the contents of the container.

❍ Standing, legs spread apart comfortably, knees slightly bent: Swing the bowl of the pelvis between the two femurs, forward and back. Initiate the movement from the sitz bones, from the pubis, then from the iliac crests. Initiate the movement from the sacrum and coccyx of the spine and allow the undulation to travel up to the skull.

❍ Standing, on one leg: move the pelvis laterally over the top of the stable femur (bending the torso side by moving at the hip joint, not at the waist). This is movement of the proximal bone, the pelvis, over the distal bone, the femur. Return to center. Change legs and move to the other side by excursioning the whole pelvis over the top of the ball of the standing leg. Return. Now stabilize both legs and bend the body forward from the hip joint (not the waist). Return to vertical. *(continues)*

✎ Draw the pelvis from the lateral view without looking at a diagram. Use your hands to find the shape of your own pelvis. Draw it from the anterior view. Compare with a diagram and draw again if necessary. Label the parts and the views.

❍ Do a dance of the pelvis: move from the arches, holes and curves of the pelvic bones; feel the weight of the pelvis and let the organs move in their bowl.
❍ Visualization: Lying in constructive rest, paint the inside of the pelvis a color of your choice. Take lots of time to feel all the surfaces of the bones.

Specifically for dancers: hip rotation

Rotation occurs at the hip, the acetabulum. Rotation, or "turn out," allows efficient movement to the side and diagonal, and enhances a frontal or "staged" view of body position for use in the theater. It also reinforces the many spirals inherent in the body, and encourages three-dimensionality in movement. In rotation, the ball of the femur moves within the acetabulum of the pelvis; the movement spirals down the axis of the leg through the foot.
❍ Standing in parallel: Move the greater trochanter of one leg back towards the ischial tuberosity of the pelvis (the back of the pelvis). Watch the effect on your knee and foot. Return to parallel by moving the greater trochanter wide to the side; this returns your leg to parallel position. Do both legs simultaneously. You are now in an open first position.
❍ Place your hands on the greater trochanters, and fan them open wide to the side as you return to parallel. Important: do not rotate the foot more than the hip. The knee faces directly over the toes, connecting the alignment from hip to foot. Plié, especially grand plié, is a continuation of rotation, not just a downward action. Because the hip is a ball and socket joint, it involves three dimensional spiral action.
❍ Try a second position grand plié. Individuals with limited rotation in standing can adjust the heels slightly forward as they plié in second position to allow deeper movement; as the knees are extended, the heels must return to standing rotated alignment.
❍ Repeat: place your hands on your thighs, thumb in front and fingers behind. Plié deeply; allow the femur to rotate in the socket so that the thumb moves to the top of your thigh, the fingers are down below. This rotation of the femur is imagetically described as "wrapping the thighs," "rotating the seam of the tights forward," or "yawning open the legs."

Exceptions are possible when working with the body. Some people have a rotation between the femur and the tibia which affects basic alignment of knees over toes. In this case, create your own working pattern, and be consistent. In dance, with exaggerated movements requiring quick, reflexive responses, efficient patterning contributes enormously to skill and safety.

ILIOPSOAS

The iliopsoas is the primary muscle of integration in the body. It connects the legs to the pelvis and to the spine. It attaches to the lumbar spine, with the crus of the diaphragm, and is affected by breathing.

The iliopsoas is composed of three muscles, each with a different function. The **psoas major** attaches at T 12 (the twelfth thoracic vertebra) and along the bodies of the five lumbar vertebrae; it then travels forward over the rim of each pubic bone and down to the legs where it inserts at the lesser trochanter of each femur, your inner thighs. Thus it spans a long distance and integrates the spine with the legs. It is involved in flexion of the hip when the spine is stable as in lifting the knee in standing posture; and in flexion of the spine when the legs are stable, as in a sit-up. The **psoas minor** originates at T 12 and attaches on the rim of each pubic bone, integrating the spine with the pelvis. It is involved in maintaining horizontal alignment of the bowl of the pelvis in standing, which keeps the organs from spilling forward against the abdominal wall. The **iliacus** attaches on the iliac crests and the inner surfaces of each ilium, travels forward over the pubic bones, and inserts on the lesser trochanter of each femur. Thus, it integrates the pelvis with the legs, providing force and endurance to flexion at the hip, such as in the kick of a football, or long distance running. All three muscles, working together, integrate the central zone of the body.

The iliopsoas spans a long, diagonal path from T 12, forward over the rims of the pubic bones, and back to the inner legs, like a sidewards V or the prow of a ship. The forward thrust of the pubic bones creates a pulley,

Finding Center

In anatomy class, I studied the sequencing of the abdominal muscles. The transversus, the interior and exterior obliques, and the rectus abdominis work collectively for front surface support. The transversus, the innermost abdominal muscle, wraps like a cummerbund around the internal organs, and initiates the sequencing for efficient movement. "Because it is so deep, you can locate it by laughing, coughing, or throwing up," my teacher said. "Take your choice ."

❖

At a summer dance workshop in Aspen, Colorado, my teacher Dena Madole demonstrated the epitome of precise, effortless movement. She had performed with the Erick Hawkins Dance Company, and she emphasized a soft, subtle spine. She insisted patiently that we initiate our movement from the pelvis – the center of our body, and referred to a muscle group called the iliopsoas. For hours each day she would reverently guide us through rocking movements of the pelvis and legs to release the grip of the thigh and abdominal muscles. "Because the iliopsoas spans from the front of the spine and the pelvis to the legs," she said, "you can let your legs dangle like tassels from your hips." She also spoke about measured energy. Facing a group of aspiring performers one evening she commented, "What I am interested in is what you do with the energy that is left when a performance is over."

**Sculpture: Michael Singer
"Lily Pond Ritual Series"**

giving added leverage to this important muscle. If we connect the span of the iliopsoas to the energy of our plumb line, our vertical axis, we have a triangle of support.

The iliopsoas works with the abdominal muscles for front surface support of the body. The iliopsoas, however, attaches along the front of the spine, allowing maximum efficiency in bone movement as well as articulation of each vertebra. The iliopsoas is important for movement initiation. The abdominals connect to the sternum, ribs and pubic bone in front of the organs, providing secondary support for the spine as well as protection and support for the organs. (The primary support for the organs should come from the stable position of the pelvic bowl through the psoas minor.) The abdominals are necessary for strength and endurance once a movement has been efficiently initiated. They also add mobility of the trunk due to their three-dimensional weaving of fibers. Thus, an efficient sit-up initiates from the psoas first, the muscle closest to the spine, then sequentially adds the layers of abdominal muscles: the transversus, rectus abdominis, and the obliques. The rectus abdominis is divided into quadrants. It is the muscle builder's rippling belly which you see displayed at the beach and is well known from emphasis in physical fitness. Our culture has invested considerable attention to the more visible abdominal muscles at the expense of the deep and highly important iliopsoas. Efficient sequencing of muscle contraction allows mobility as well as strength in the spine.

How can you tell if you are using the iliopsoas? The belly will hollow, rather than bulge forward as you contract; the iliopsoas pulls towards the spine rather than away. You will also be able to move sequentially through each vertebra rather than the spine moving stiffly or in blocks. Sit-ups initiated from the powerful and visible rectus abdominis, can cause rigidity of the back resulting in lower back pain. ❖

*I took a workshop with Chungliang Al Huang, a tai chi master and author of the book, **Embrace Tiger, Return to Mountain.** He focused our dance movements around a mobile center. It felt great. Years later, I studied with him in China. Let the pelvis be free. Don't stand with your spine rigid like a pole, he encouraged his aspiring but tense tai chi students. Your center is the source of your energy, your tant'ien.*

❖

I was teaching a student skilled in the martial arts. Working with centering was part of his basic training, but his tight, bulky muscles restricted his movement. He became obsessed with learning about his iliopsoas. First his tight lower back and gluteal muscles released as he began initiating movement from his front surface. Eventually his abdominal and thigh muscles began to relax as the deeper iliopsoas muscles integrated his spine and legs. His body shape changed. What remained consistent was the wide-eyed grin on his face. He was an architecture major. "This makes sense," he said.

Anterior view of iliopsoas and deep muscles of trunk and hip

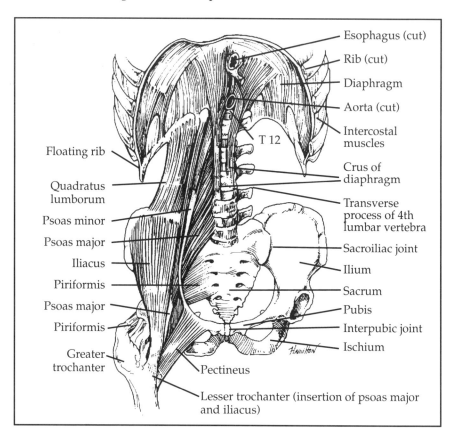

Sit backs
15 minutes

Seated on the floor, relaxed spine, knees bent in front of your chest:
○ Place your hands under your thighs for support. Looking at your belly, relax the muscles and feel the center hollow, knees bent, feet flat on the floor. Image that you are lying in a hammock to facilitate the release; or that there is a cat in your lap, gently pawing at your belly. Gradually scoop the pelvis, as though someone is pulling your tail up between your legs. Image energy circling down the back and up the front of your body. As you do this, feel the fifth lumbar vertebra touch the floor. Gradually lower each vertebra to the ground by articulating the pelvis rather than by lowering your ribs. Continue to watch your belly and to breathe freely with each movement. If the rectus abdominis bulges forward, pause in your movement and try to relax your center so the deep iliopsoas can do the work. Sway the torso side to side gently to encourage release. You must use your arms for support; too much weight necessitates abdominal assistance.
○ Lay the entire spine out on the floor, including the cervical vertebrae, like a bicycle chain. Release the weight of the skull into the earth.
○ Take a few deep breaths, feeling the belly move. Begin the sit-up portion of the exercise by curling the skull forward (the first body weight). Then roll through each thoracic vertebra using your arms on the floor beside you. Rest in "beach-lying position," propping yourself up by resting on your forearms and elbows, legs slightly bent as you observe your abdominal muscles. Continue the sit-up, pausing whenever the rectus abdominis pops forward. Use your hands on the back of the thighs and sway slightly side to side to encourage abdominal release in this "hammock" posture. Articulate the entire spine as you roll forward, then open the knees slightly and hang the torso between the legs. Place your hands on your breathing spot (on your lower back). Feel your breath massaging these often tight muscles. Repeat, moving as slowly as you can. This exercise often takes several weeks to accomplish effectively.
Be patient. Muscle sequencing requires letting go and relearning a pathway in your body. It is like being shown a shortcut home from school. You might have walked one route for years, but once you learn the new one, it is there for your use.

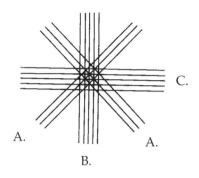

Energy down the back, up the front

Abdominal muscles, fiber directions:
A. Internal and external obliques
B. Rectus abdominis
C. Transversus abdominis (innermost, next to organs)

Sit backs, with partner
10 minutes each person

One person standing, the other seated on the floor facing each other:
○ Standing person places his/her feet outside the feet of the seated person; both have slightly bent knees. Clasp wrists firmly.
○ Standing person curve your lower back for support.
○ Take the full weight of the seated person through your skeletal structure. Begin lowering your partner vertebra by vertebra to the floor, as your partner curves their pelvis forward.
○ Seated person allow yourself to "be lowered" without using the rectus abdominis.
○ Standing person, observe your partner's belly as you lower him/her to the floor. When the rectus abdominis pops out, pause or gently jiggle the body through your arms to encourage release.*
Don't judge or command; you are encouraging release, not tension!
○ Lay the entire spine slowly on the floor, bending your legs more deeply as you do so for counterbalance and protection of your spine. Breathe.
○ The person lying on the floor begin curving the skull forward to initiate the roll up; allow the rest of the body weight to be supported by your partner. Concentrate on articulating through the spine with the deep iliopsoas muscle, releasing the outer abdominals.
○ In seated position, standing person stretch your partner's torso forward over their legs. Repeat.
○ Talk about your experience. Change partners.

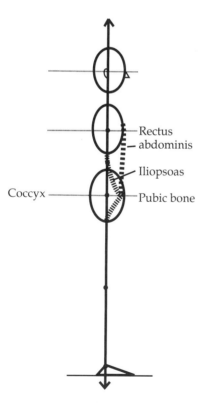

Rectus abdominis
iliopsoas
Coccyx
Pubic bone

Rectus abdominis and iliopsoas in balanced alignment with three body weights

* The abdominal muscles are important for strength, endurance and directional mobility of the torso. We are working to locate the deeper iliopsoas for movement initiation and strengthen it for spinal articulation. Then we can add the abdominals.

Front surface alignment: "bellital" alignment*
15 minutes

❖ Lying on the floor, belly down:
○ Be aware of your forehead on the floor. Let your brain, throat and neck relax towards gravity.
○ Be aware of your sternum on the floor. Let the weight of your lungs and heart rest on the sternum and ribs.
○ Feel the pubic bones and belly on the floor. Allow the organs to rest supported by the abdominal sheath and the pubic bones.
○ Connect your entire front surface: Trace from the pubic bone, to the rectus abdominis, the bottom of the ribs and the sternum, to the hyoid bone, the mouth and the forehead.
○ Be aware of the knees and toes on the floor. Allow the muscles of the legs to relax.
○ Flex the toes and begin gently rocking through the front surface supports of the body. Constructive rest brings awareness to back surface support; bellital alignment brings awareness to front surface support.

❖ Image a line painted down your forehead, nose, chin, front of your throat, sternum, to belly button and pubic bone. Sequentially peel this painted line off of the floor. Use your forearms and hands as needed for support. Feel the connection from forehead to pubic bone. Return the torso to the floor, elongating the spine.

❖ Standing: Image a forearm on the front of your body, the elbow on the pubic bone and the palm of the hand on your sternum. Image a forearm on the back of your body, the elbow on your sacrum and the hand behind the heart, connecting the body weights. See these forearms as two parallel supports. Feel the diaphragm free to move up and down inside. Image the diaphragm as a piston inside a cylinder.

* Term developed by Caryn McHose

Bench: Kristina Madsen

FEMUR:
A Leg to Stand On

The femur is an active bone. It has a curved "ball" at each hip plus two curved "ball" surfaces at each knee, creating a high potential for movement. Hence the amount of musculature at the hip and thigh! The hip is a ball and socket joint with three-dimensional action and the knee is a hinge joint with two dimensional action; the hip serves to establish direction of movement while the knee determines the range of that motion.* The femur is the longest bone in the body and an important site for red blood cell production. It has a "neck" which extends the shaft of the bone laterally away from the pelvis so that the bowl can swing freely between the legs. It is very important to image the femur bone going away from the pelvis, then descending diagonally back to join the plumb line made by the hip socket, the knee, and the ankle. In vertical posture, our parallel leg alignment is determined from this plumb starting at the acetabulum. The greater trochanter is the knob you feel on the outside of the leg by the pelvis; it is the protuberance which gets sore when you are lying on your side on a hard floor. Touch this place on your leg as you sit. Directly opposite the greater trochanter, on the inside of the femur – the inner thigh – is the lesser trochanter, the attachment site for the iliopsoas muscle. Clear imaging of the space between the pelvis and the greater trochanter encourages flexibility at the hip and consequently reduces knee strain. ❖

*See Arthrometric Model, p. 114.

Holding

I spent a week in an apartment in New York alone. Each day I would lie on the floor and do bodywork before venturing out into the city to explore. It seemed ironic, to be in a huge city and do such private work. I began working with the hip socket, circling each leg slowly, feeling the reactions as nerves and muscles released. One evening my leg began jerking uncontrollably. Somehow, I was more interested than afraid. My mind observed as I felt my whole body thrashed around from the vibrations in my leg. Thoughts passed in and out as the body did its own release: memories of crossing legs properly as a young lady should, daily classes since childhood developing rotation at the hips for classical ballet (with hip sockets not quite shaped for that activity), deep wide "second positions" and hundreds of pelvic contractions in modern dance, sexual stories – warm and safe, cold and confusing. I started crying, more from relief than from pain or fear. When that side was exhausted, I slowly began the other side. A similar process occurred: the circling, then jerking, vibration, and extension out through the foot until the whole body was shaking. On and on the mind went in partnership with the body, on a floor in New York with people going about their multitude of activities. I repeated the work for the next few days, with less dramatic responses but still with a sense of releasing old tensions and of recognizing all that is stored in hips and thighs.

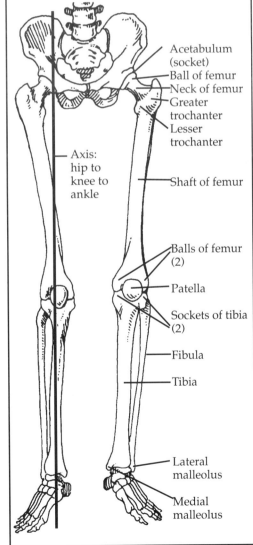

Acetabulum
(socket)
Ball of femur
Neck of femur
Greater
trochanter
Lesser
trochanter

Axis:
hip to
knee to
ankle

Shaft of femur

Balls of femur
(2)

Patella

Sockets of tibia
(2)

Fibula

Tibia

Lateral
malleolus

Medial
malleolus

Pelvis and legs: anterior view

Hip circles
20 minutes

Lying in constructive rest:
○ Lift the right leg until the femur is perpendicular to the earth, the knee relaxed and the foot off the floor; stabilize the left leg by allowing weight to go firmly into the foot to the floor, knee flexed.
○ Begin to trace tiny circles with the right knee on the ceiling – as though you were drawing with a pencil attached to your knee. Move very slowly, so you are working underneath the big muscle groups and stimulating the deep rotators and the joint capsule. You may experience nervous system "jerks" as the big muscles let go and the deeper muscles are stimulated – allow these jerks.
○ Move slowly. Gradually increase the size of the circle until it is the range of your hip joint. If you are very relaxed, you may be able to trace the rim of your hip socket with the neck of the femur. Don't go beyond your comfort level at this slow speed.
○ Reverse the direction of the circles until you are again making tiny circles around the perpendicular axis of the leg.
○ Place the foot on the floor and do the other leg. Remember the slower you work, the deeper the release in this particular exercise.
○ Repeat at another time with a partner holding and circling your leg so you can release all muscles except the stabilizing side. Again, go slowly! Support the knee carefully from underneath and hold the heel.

Femur: tracing the bone
5 minutes

○ In seated position, place your fingers on the greater trochanter. It is the knob on the outside, top, of the leg. Place your thumb on the inside top of the femur, the lesser trochanter. Massage through the muscles of the inner thigh to feel the width of the bone; the inner portion of the thigh is all soft tissue such as muscle and fascia.
○ Keeping your fingers in this relationship, pincer (squeeze) the fingers along the bone to trace the length of the femur. Remember that the femur makes a diagonal from the hip to the knee. Carefully trace the bottom of the femur and its connection to the tibia. Feel the width of the knee. Do not apply pressure to the patella, or the "knee cap." It is a free-floating bone.

Thigh rolls
5 minutes

○ Placing your hand on the thigh, gently rotate the femur bone in one direction as you roll the thigh muscles with your hands in the opposite direction. Reverse. Feel the sensation of bone moving within muscle.

Whole leg: toothpaste tube
5 minutes

With a partner: Stand vertically; have your partner squeeze the soft tissues of the leg from the thigh all the way down to the foot, like squeezing a toothpaste tube. Do both legs. Change partners. Feel the fluids move through the tissues.

Swinging the leg
5 minutes

Standing, both feet on the floor, knees slightly bent, legs spread apart for a wide base:
○ Stabilize the left leg. Swing the right leg across the body initiating from the right foot. Return to standing position.
○ Repeat the swing initiating from:
    the right greater trochanter,
    the right acetabulum,
    the right sacroiliac joint,
    the left sacroiliac joint down to the left foot.
○ Keep both legs slightly bent. Repeat swings with left leg.

Sculpture: Michael Singer
"Cloud Hands Ritual Series"

KNEE

The knee transitions the weight from the torso to the ground. Each knee is composed of two balls and two sockets. The patella, our "knee-cap" is a free-floating bone which slides in the groove between the two balls of the femur. It is inserted in the quadriceps tendon which connects the thigh muscles to the lower leg. This continuous strip of connective tissue is called the quadriceps tendon above the knee, and the patellar ligament below the knee. The patella functions to protect the knee from impact, like a shield, as well as to create a fulcrum for additional leverage for the quadriceps muscle. The patella is *not* a weight-bearing bone in standing position. Two discs, or menisci, cushion the transfer of weight between the balls and sockets within the knee. These discs are thick at the outer edges and thin in their centers to deepen the sockets for stability of the femur. Cartilage is attached to the bone ends for added resiliency. The entire structure is surrounded by a fibrous joint capsule filled with synovial fluid to nourish the bones and discs and to reduce friction. The fibrous capsule is continuous with the ligaments and tendons around it. The protective ligaments of the knee provide stability as well as mobility at the joint. These include the anterior and posterior cruciate ligaments which cross within the joint, interlacing the two balls and the two sockets to minimize internal slippage; and the tibial and fibular collateral ligaments which stabilize the sides of the knee and give lateral support. The direction of the attachment of ligaments to bones helps to pattern muscle contraction and equalize the angles of stress at the joint. In general, the type of connective tissue which forms ligaments, tendons and joint capsules does not recontract and should not be stretched.

The knee is a hinge joint, with the two balls and two sockets allowing two-dimensional movement at the joint: flexion and extension. When the knee is flexed, or bent, there is slight rotational mobility. This, however, can stretch the ligaments and is to be kept to a minimum. In efficient postural alignment, the weight of the body transfers directly through the center of the joint, not the front or the back. Stand and feel your own knees; become aware of the sensation of bone standing on bone with minimal muscular involvement. If the knee pushes backwards behind your plumb line, you have gone past the place of bone alignment and are "hanging in your ligaments." This is often accompanied by a forward thrust of the pelvis, using ligament support at the hip as well. Bring your bones into balance; it may feel like you are bending your knees when they are actually straight. Once ligaments are stretched beyond sufficient support for postural alignment, you must rely on muscles to do the work.

Connections

When I was first teaching Anatomy and Kinesiology for Dance at Mount Holyoke College, I took a student with a knee injury to see Bonnie Bainbridge Cohen. The student was a dancer and was considering knee surgery. Bonnie looked at the alignment of the knees, looked at the young student and said: "Once you have surgery, you can no longer rely on the strength of the evolutionary line. You have to work within the limitations of this lifetime." And we left.

Again, in postural alignment, we want the weight to pass clearly through the center of the bone ends, so that the femur is in vertical alignment with the tibia. In movement, we want even and equal distribution of weight throughout the joint. In neither instance does weight pass through the patella.

In flexion and extension at the hinge joint of the knee, the two sockets excursion around the two balls, and the balls move evenly within the sockets. The joint should not hinge "open," separating the bone ends. Instead, as the knee bends, the back of the femur becomes the bottom,

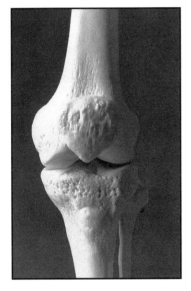

Knee: balls of femur, sockets of tibia, and patella

Illustration: medial/posterior view of right knee and ligaments in extension and in flexion

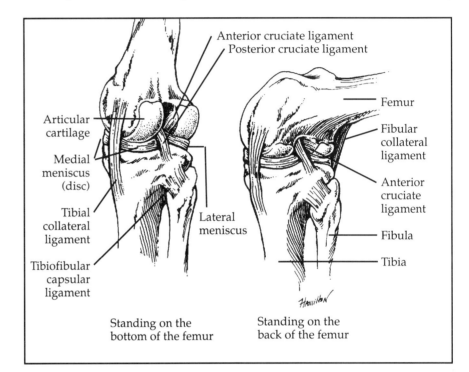

Anterior cruciate ligament
Posterior cruciate ligament

Articular cartilage

Medial meniscus (disc)

Tibial collateral ligament

Tibiofibular capsular ligament

Lateral meniscus

Femur

Fibular collateral ligament

Anterior cruciate ligament

Fibula

Tibia

Standing on the bottom of the femur

Standing on the back of the femur

articulating with the sockets of the tibia; and the bottom of the femur becomes the "front" of the knee, protected by the patella. As we extend the leg, the bottom of the femur (now the front of the knee), returns to its position as the weight bearing surface with the tibia. Feel this in your own leg. If you are seated in a chair, your leg is flexed at the knee. The bottom of your femur is exposed, protected by the patella; the back of your femur is in contact with the two sockets of the tibia. Extend one leg so that it is horizontal to the floor in front of you; now the back of the femur is exposed and the bottom of the femur is in contact with the tibia.

The discs provide "transition" between the femur and the tibia, cushion the bone ends, and absorb impact. As we stand and move, the discs transfer the weight through the joint to the lower leg and to the earth; holding at any joint transfers the function of that joint to the one above or below. For example, holding of the knee affects the hip and ankle; tightness at the hip affects the spine and knee. The joints are the site of movement for the skeleton; the alignment of one affects the whole. ❖

Feeling the two balls and two sockets of each knee

○ **Seated:** Trace the knee joint with your hands. Feel its width and its depth. Feel for the space where the bone ends meet. This is the location of each of the discs – the medial and lateral menisci which cushion the bone ends. Locate the patella, the free-floating bone which slides in the groove between the balls of the femur. Continue from the patella down to the bony knob on the front of your shin. This is the insertion site of the patellar ligament which connects your quadricep muscles to your lower leg. Place your fingers on each side of the knee. Use your thumbs on the femur and the fingers on the tibia. This is the site of the lateral collateral ligaments which protect the knee from the sides. Feel the cord-like tendons which run along the back of the knee. These are the tendons of the hamstring muscles.

○ **Seated in a chair:** Place your forearms along the sides of one femur; let each fist represent a ball of the knee joint. Stabilize your arms, and extend your lower leg. Imagine the movement of the sockets of the tibia around and under the balls. Reverse. You are moving the distal bone on a stable proximal bone.

○ **Standing in front of a chair:** Begin to sit. Focus on the movement of your knees. As you lower your weight, imagine the balls of the femur rotating within the sockets of the stable tibia. Stand. You are moving the proximal bone on a stable distal bone.

○ **Walking:** Focus on the lifted leg. Feel the tibia swing forward under the femur. As you reach with your heel, the sockets move under the balls. Focus on the leg in contact with the floor. Be sure the knee is relaxed to transfer the weight of your torso down through the leg to the floor. Walk gripping your knees; walk relaxing your knees, feeling the easy articulation of the bones.

○ **Standing in parallel:** Check hip, knee, ankle alignment. Plié, bending the hip, then the knee, then the ankle equally. Be sure the knees go directly over the second toes as you bend. Keep the spine on plumb line. Straighten the legs, from the ankle, the knee, the hip sequentially. Feel how they work together. Repeat, focusing on the change of weight from the bottom of the femur to the back of the femur as you bend; from the back of the femur to the bottom of the femur as you extend.

○ **Standing:** Rock your plumb line back on your heels, keeping your knees straight. Feel the reflex in your hip sockets to open the feet to help you catch your balance. Follow this impulse: Rock back on your heels, reflex in the hip to open the feet and toes, find balance in this open position. This is called *(continues)* "first position" in dance. Heels facing, toes open. Plié, directing the knees over your second toes. Extend your legs and rock back on your heels to return to parallel. Feel how change of direction comes from the hip joint, not from the knees.

❍ **Standing:** Imagine stepping on something sharp. Really feel it so you trigger the reflex to withdraw your foot quickly from the floor. Or imagine a hot floor, or a piece of glass. Play with tricking yourself into a hip reflex pattern (called flexor withdrawal). This pattern underlies deep flexion at the hip, knee and ankle and utilizes the iliopsoas muscle. It is the most efficient muscle patterning for lifting the leg and is useful for establishing hip, knee, ankle alignment with efficient muscle sequencing.

Caring for your knees

- Minimize impact on the front of the knee.
- Check hip, knee, ankle alignment in all positions.
- Learn the structure of the joint; visualize it clearly.
- Stand with the weight through the center of the joint.
- Massage the thigh muscles to equalize pull on the patella and the joint.
- Listen to pain at a joint; never stretch ligaments and tendons.
- Balance the use of the front and back surfaces of the legs in walking.
- Equalize use of hip, knee, ankle in flexion and extension.
- Know that the knee is a transition between the hip and the ankle.
- Relate pain at the knee to alignment of the whole leg and body.

TIBIA AND FIBULA:
The Lower Leg

The tibia is the straightest bone in the body. It is a receiving bone, with two cupped surfaces to connect with the two balls of the femur above. Below, it creates an archway with the fibula to receive the top bone of the ankle, the talus. Thus, its role is to transfer weight from the femur to the foot. The tibia is braced laterally by its companion bone the fibula; the fibula does not articulate with the femur at the knee. At the ankle, the tibia creates the medial two-thirds of the socket, and the fibula the lateral one-third. The talus bone of the ankle excursions forward and backward in the

Below the Knees

When I was twelve years old and in sixth grade at a country school in Illinois, I was in stiff competition with a boy for who was the smartest student in class. We sat across the aisle from each other. I would be at my desk, swinging my feet (which never touched the ground in the tall chairs throughout grade school). I watched him receive his 100% spelling tests, and would wonder if it was easy for him or if he worked hard to learn. He was always good and very interested in pleasing the teacher. One day on the playground, I went up to him – I don't remember why – and kicked him as hard as I could in his shins. I ran back in the building to the top of the stairs and looked out the window. There he was, crying in pain with all the other children laughing. I was frozen, watching. No one ever said a word about it to teachers or parents. But he stopped getting good grades, and I moved away.

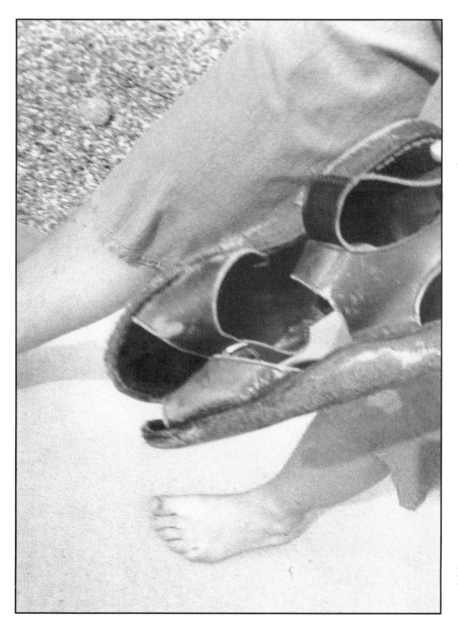

Photograph: Bill Arnold
"Feet and Sandals"

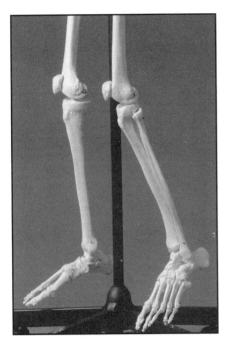
Tibia and fibula

curved archway as the foot flexes and extends in movements such as walking. The ankle is considered a hinge joint, with two-dimensional movement. The talus also works as a gliding joint, articulating as one of the seven tarsal bones forming the arch of the foot with the metatarsals. Their collective action provides the capacity for three-dimensional mobility of the foot; warm-ups which locate "circling the ankle" at the hinge joint between the tibia, fibula, and the talus, risk stretching ligaments and losing stability at the ankle. Both the tibia and the fibula rotate slightly around their own axis in relation to movement of the talus in locomotor activity. They are woven together by the interosseous membrane, an interlacing of fascia which gives lightweight support between the two bones.

"Shin splints" is a common term for many types of stress of the muscle, bone, and interosseous membranes of the lower leg. Allowing the subtle rotation of the tibia and fibula in vigorous activity massages the interosseous membrane and keeps fluids moving through these tissues. Exercises which repeat flexion and extension of the muscles of the lower leg over a long period of time can fatigue the muscles beyond resiliency and cause tearing or fracture of tissues. Balancing reflexive muscle contraction with generalized movements, rest, and muscle stretching (not tendon stretches) reduces the likelihood of injury; massaging muscles before and after use, allowing rotation of the bones of the lower legs in movement, and equalizing flexion at the hip, knee, and ankle joints reduces stress. Efficient postural alignment underlies ease of movement throughout the body. One task is to listen when the body speaks signs of fatigue, and respond before injury occurs. ❖

Tracing the tibia and fibula

Seated on the floor: Trace the tibia of your right leg, from the knee to the ankle. Begin by feeling the two sockets and their connection to the balls of the femur.

◯ Continue down the tibia (your "shin" bone) feeling the width as well as the length of the bone.

◯ When you arrive at the ankle, trace the bottom of the tibia, the inner malleolus, and continue around to feel the cupped surface of the ankle joint. Here you will meet the fibula, the outer malleolus, which forms the lateral one-third of the joint. Pincer up the outer malleolus and continue along the fibula until you reach its connection to the tibia, below and lateral to the knee. Feel how the fibula supports the knee and the ankle.

◯ Move your fingers to the space between the two bones and massage down the interosseous membrane which weaves the bones together. This may be quite sore for some people; use a firm but sensitive touch. Repeat with the other leg.

With a partner: The person touching is seated on the floor, and the person being touched lies in constructive rest. Supporting the back of the knee and holding the heel with your hands, carefully lift your partner's leg and drape it over your leg, the knee supported on your thigh. Always be sure that the leg is aligned between hip, knee, and ankle; no twisting at the joint.

Right foot, pointed; right foot, flexed

Rotating the tibia and fibula

On your own: Place your forearms parallel to the tibia on one side, and the fibula on the other. Your elbows will match the knee and the fists of your hand the two malleoli.

○ Flex your ankle, drawing the toes towards the front of your shins, and rotate your arms around themselves so the back of the arms face each other. This is the motion, in exaggerated form, of the tibia and fibula as you flex your ankle; the bones rotate slightly, around their axes, towards the front to allow the talus bone of the ankle to move to the back.

○ Now, point your foot, extending the toes and foot towards the floor, and rotate your forearms so the belly of the arm comes forward. This is similar to the rotation which happens in the tibia and the fibula to allow the talus to move to the front of the ankle for support. Do this several times to clarify the image. As the foot flexes, the bones rotate around their axes to the front to allow the talus to move back; as the foot points, the bones rotate out allowing the talus to move to the front of the ankle. Touch the talus as the foot points; feel it disappear behind the tendons as you flex.

○ Repeat on both legs.

○ Stand. Bend your knees (we will refer to this as a plié – a French term in dance meaning to fold), flexing evenly at your hip, knee, and ankle while keeping the spine on vertical.

○ Extend your knees until the femur balances vertically on the tibia.

○ Fall forward with your plumb line until you go up to balance on your toes (a relevé in dance terminology).

○ Repeat, imaging the movement of the talus back as you plié, and the rotation of the tibia and fibula outward and the talus moving forward as you relevé.

○ Move your arms with the rotation like the legs to help clarify the movement. Repeat slowly until this pattern is clear. The rotation in the tibia and fibula is quite small. It makes a tremendous difference, however, in the efficiency of movement in the lower leg. This natural rotation helps in absorbing the shock of impact in running and other high impact sports; and it massages the interosseous membrane, pumping the fluids, and bringing oxygen, removing waste in the hard working leg muscles.

○ Take a walk and feel the movement in the lower leg as you reach forward with your heel (flexion of the ankle) and push off behind with your toes (extension of the ankle).

FEET

The feet are our base of support and our connection to the ground. They contain many nerve endings for sensing and responding. Shoes, concrete, flat sidewalks, elevators and escalators encourage most of us to lose the articulate sensitivity which our feet possess. In the insect world, the feet are tongues for the moth! Our feet receive and express our connection to our environment. They spread the weight of the body to the ground, give stability and mobility to our base, absorb the shock and impact of motion, and give articulation and refinement to our movement and gesture. Our feet constantly inform us of the stability of our base. Whether we are on a boat or on land, on sand or on rocks, safe or uncertain, they respond accordingly.

There are seven tarsal bones in the ankle. They allow circular movement with maximum stability and a minimum of musculature (imagine if our hip muscles were at our feet!). Each irregular bone surface slides minutely over the next to collectively provide range of motion. The uppermost ankle bone, the talus, is braced between the two bones of the lower leg, the tibia and fibula. It is allowed to move forward and backward with each step, like ice swinging between ice tongs, adding to the overall leverage of our striding gait, and assisting the tibia and fibula in their slight rotation to reduce impact at the knee. The talus feeds directly forward into the navicular bone, the three cuneiform bones, the first three metatarsals, and then to the first, second and third toes. Collectively, this structure creates the "ankle foot," levering support from the first three toes to the front of the ankle. Returning to the talus, the weight also transfers posteriorly to the calcaneus or "heel" bone. The talo-calcaneo joint has three roughly curved surfaces, corresponding to the three planes of movement, and allowing for subtle adaptation to uneven terrain. Stress is absorbed through this joint in small doses before traveling up to the ankle, knee and pelvis. The calcaneus also connects forward to the cuboid bone, the fourth and fifth metatarsals, and the fourth and fifth toes (the big toe has two bones forming hinge joints, whereas all the other toes have three bones each: a total of 14 phalanges per foot). This connection from calcaneus to cuboid to metatarsals to toes is called the "heel foot," giving support to the posterior surface of the ankle. Thus, in standing and in movement, when all five toes are engaged, the ankle receives support at both the front and back surfaces. All of the bones of the foot work together to create the horizontal and longitudinal arches, which cushion and give spring to our step. The levering from big toe to heel is also a vital part of our powerful walking stride.

**Costume Design: Kristen Kagan
For choreography and performance
by Andrea Olsen**

Slow Down

I was advising a student who was determined to be a professional dancer. She had attended a performing arts high school, and had distorted her body in every conceivable way to force it into a "perfect" dancer's mold. Stress was part of her working process. She broke her foot (fifth metatarsal) four weeks before her senior concert. Rather than cancel her performance, she made new work based on her injury. She began a dance in a chair, following authentic impulses in her body. Her work changed completely. Because she couldn't stand, couldn't rush, she had to listen to her body. Movement that had been superficial or imitative, was filled with personal feeling and vulnerability. Slowing down allowed her to move.

Foot: left lateral view in illustration
Foot from above

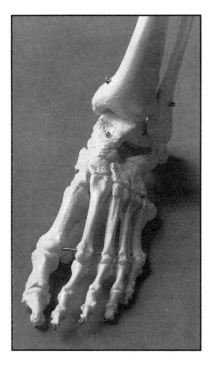

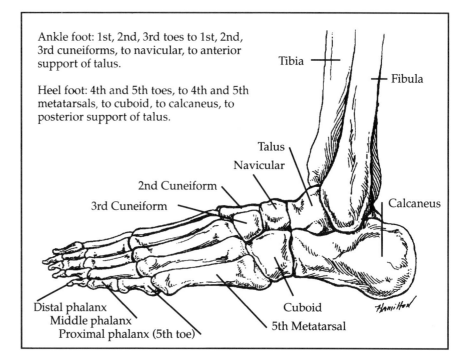

Ankle foot: 1st, 2nd, 3rd toes to 1st, 2nd, 3rd cuneiforms, to navicular, to anterior support of talus.

Heel foot: 4th and 5th toes, to 4th and 5th metatarsals, to cuboid, to calcaneus, to posterior support of talus.

Tibia
Fibula
Talus
Navicular
2nd Cuneiform
3rd Cuneiform
Calcaneus
Distal phalanx
Middle phalanx
Proximal phalanx (5th toe)
Cuboid
5th Metatarsal

The foot has a role in communication and emotional stability as well as in functional support. Instability of the base of support is disorienting, as anyone who has had a broken toe or foot bone will attest. Proper care and attention to the feet can affect such far reaching problems as pain at the knee, hip, sacroiliac joint, or spine, organ stress, headaches, and general malaise. Massage the feet, walk barefoot on a variety of surfaces, sensitize the skin by gentle brushing, circle the toes and notice the difference! ❖

Foot massage
15 minutes each foot

Seated, place one foot on top of the other, or find a comfortable position for working with the foot:

○ Circle the little toe with your hand. Move all three phalanges of the toe collectively in their socket, like stirring a bowl with a big spoon. Move slowly and focus on the bones rather than muscle. Continue through all five toes.

○ When you have finished with the big toe, walk your fingers along the first metatarsal. Then pulse your fingers in the soft tissue between each of the five metatarsals, placing the fingers on the top and the thumb on the bottom of the foot, like pincers. Use both hands if necessary; you are encouraging equal and even space between the bones.

○ At the fifth metatarsal, walk your fingers up to the cuboid bone, which is a squarish bone above the fourth and fifth metatarsal.

○ Continue across the arch of the foot and feel the third, second, and first cuneiforms. Above these, trace the shape of the navicular bone, the center of the arch, and its connection to the talus of the ankle. This grouping of bones, the first second and third toes, their metatarsals and cuneiforms, into the navicular and the talus is called the "ankle foot" because it supports the ankle from the front.

○ Move your fingers back to the cuboid bone, on the little toe side of the foot. Move from there to feel the large calcaneus bone which continues all the way to become your heel.

○ Trace the connection between the calcaneus and the talus. The levering from the fourth and fifth toes to the cuboid bone, to the calcaneus and up into the bottom and back of the talus is called your "heel foot" because it gives support to the ankle from the back.

○ Now, hold the entire foot in your hands and massage it through the soft tissues. What feels good is what you want to do! Take your thumbs and press on the bottom of the foot from your heel to the toes, along the longitudinal arch.

(continues)

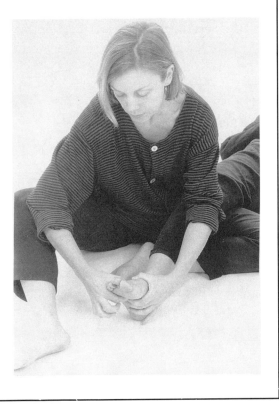

Note:

The more stable our base, the more relaxed our central joints; as our base is secure, the center can be free from holding and ready for movement. Observe.

○ Next, place your thumbs firmly on the calcaneus and massage with one stroke, from your heel to the toes.

○ Hold your calcaneus and circle the toes by moving the heel; observe the movement through the tarsals, metatarsals, and toes.

○ Place one arm under the back of your leg; let the foreleg relax and fold over your arm. Gently jiggle the leg in the hip socket.

○ Place your foot close to your pelvis and relax the other leg so your hips are squarely on the floor. Fold your torso forward towards your feet, for a gentle stretch of hip and back. As you hang in this position, place your hands on your lower back area – your "breathing spot" (between pelvis and ribs). Feel the skin and muscles of the lower back moving under your hands with each breath.

○ When you are ready to come up, press your sitz bones into the floor. Allow your weight to pour down into your pelvis to bring the spine sequentially up to vertical; press down to come up.

○ Open your legs and hang over each one. See if there is any difference in the amount of stretch. Move through the transition from floor to standing. Feel your feet and legs.

○ Do the other side. Stand.

○ Walk, reaching with the heel, pushing with the toes.

○ Run, feeling the articulateness of the feet.

Foot massage with a partner: same position as lower leg massage.

Integration of the pelvic girdle: feet to pelvis
5 minutes

Seated on the floor, soles of the feet pressed together, spine and head vertical:

○ Feel or image a connection between your big toes and your pubic symphysis. Add pressure between the big toes and gently rock on the pelvis to feel how the bones relate. Relax.

○ Feel/image a connection between your second toes and your ischial tuberosities (sitz bones). Relax.

○ Feel/image a connection between your third toes and your hip sockets. Relax.

○ Feel/image a connection between your fourth toes and the back of your pelvis: the sciatic notch and the sacroiliac joint of each pelvic half. Relax.

○ Feel/image a connection between your fifth toes and your iliac crests.

○ Feel/image a connection between the soles of your feet and the hole in your pelvis.

○ Stand and feel the connectedness of your feet to your pelvis.

Integration of the pelvic girdle: chair work

Transition from sitting to standing; collective action of the joints:
Seated on a straight-backed chair, feet in parallel flat on the floor, arms dropped by your sides on their plumb: Check parallel alignment of the legs by tracing a line from your second toe, up the center of the ankle, to the center of the knee, to the hip socket of each leg.

○ Pressing the left leg into the floor and the sitz bones into the chair for support, lift the right leg from your iliopsoas. Swing the leg open to the right like a barn door opening. Swing it back to parallel and place the foot on the floor. Repeat with left leg.

○ Lift both legs, open side and return to center. Use the sitz bones as your feet.

○ Stabilizing the left leg, lift and open the right leg to the side. Touch your right greater trochanter with your hand and feel the knob move towards and away from your pelvis as you open and close. Now, rotate the greater trochanter down towards the chair and let your foot come up, like twisting a screwdriver. Bring your leg back to parallel and place on the floor. Repeat with left leg.

○ Lift the right leg and begin moving the leg by following your little toe. Then your fourth toe, then the third, second, and finally let your leg move by following the big toe. Return to parallel and place on the floor. Repeat with left leg.

○ Open both legs to wide position, put your feet on the floor, and stand by pressing your weight down into your feet to stand up. Sit and return to parallel.

○ Stand from parallel. Check hip, knee, ankle alignment as you move. Feel your way of moving from seated to standing. Think how many times we do this in a day!

○ Sit again. This time, lead by reaching forward with your conehead, and allowing one foot to be slightly in front of the other for leverage. Sit down by reaching with the tail. Repeat this several times. See how efficiently you can lever your weight to standing by reaching with the head, shifting the body weights, and pushing down into the floor to stand, and then to walk.

○ Lying on the smooth floor near a wall in constructive rest: Lift your feet and place them against the wall. Be sure there is a clear path behind your head; then push yourself away from the wall by levering through all your joints. Repeat several times.

Then stand and jump. Push against the floor to go up; absorb the shock of the landing by sequentially flexing the toes, ankles, knees, and hips.

Specifically for dancers: relevé

❖ *Relevé*, or rising to the ball of the foot, is both a process and a position. It is an extension of the length of the leg and the height of the body by pushing down into the earth to rise. In the process of relevé, you establish parallel or rotated alignment at the hip, through the axis of the leg, down the center of the ankle. Twisting, or falling to the inside or outside of the ankle or foot (rotation, pronation or supination) is dangerous in weight bearing. In can be chosen as a gestural touch when the foot is in the air, such as a beveled ankle, but the structural alignment needs to be established before contact with the floor.

○ Stand in parallel facing a barre or chair for support through your hands.

○ Demi plié. Flex or bend the hip, knee and ankle equally; feel the depth of your plié through the muscles, without pushing into bone limit, which can injure joints. Observe the alignment of hip directing the knee over the ankle and toes.

○ To extend, push your feet into the earth, using the heel, ball and toes equally until the bones are vertically aligned; the knees are "straight" but not locked.

❖ Images are used in classes to facilitate the illusion of lightness in relevé, such as "lifting from the top of the head." The mechanics, however, involve pressing down into the ground to lever the body up, as the center of gravity shifts subtly forward over the ball of the foot.

○ Move slowly and look down to observe the energy line traveling in parallel position from the toes through the ankle and knee to the hip.

○ Repeat slowly several times. Establish a clear, consistent path from the second toe through the talus i.e., no wobbling side to side. This is a good exercise to do immediately after a minor strain of knee or foot; it realigns the axis of the joints and removes confusion in nerve pathways. Do not use if painful. When the legs are rotated, as in first position, the process is similar. Increased rotation at the hip is added for stability in the shift of weight in relevé.

○ Feel your center of gravity return over the center of your base in standing position.

❖ *Balance in relevé* is an energized position with the bones giving primary support. The toes and the metatarsal ends create a platform for relevé. Too much forward arch reduces stability to the front (overstretched ligaments), too little arch keeps you from arriving securely at the platform.

○ Feel your own relevé. As the tarsal bones align vertically, the talus is the connection to the tibia and fibula. If the talus presses forward, you lose stability; if it stays too far back there is insufficient arch for bone balance on vertical. Strength-building occurs during the process phase of relevé; sensitivity to bone balance occurs in the positioning. Remember to feel support from the heel foot and the ankle foot for back and front leverage in relevé. ○ Lower, feeling your heels connect to the floor; the image of "kissing the heels to the floor" helps sensitize your awareness to this important phase of alignment.

❖ Repeated pliés and relevés, flexion and extension in quick succession at a high speed, can cause contracture of the muscles, a state of overfatigue resulting in loss of responsiveness. Jumping becomes dangerous. Often we warm-up so extensively, that our muscles are already close to contracture before we begin rigorous dancing. To avoid this problem, include muscle stretch and full bodied movements emphasizing varying qualities to allow resiliency in muscle tone. Observe the effect of warm-ups and classes on your body responsiveness.

Efficient performance preparation includes:

• Alignment sensitivity
• Warming of joints and tissues
• Full body integration and concentration

One well-known modern dancer accomplishes all of these tasks by doing a single grand plié before performing.

Drawing: Harriet Brickman
"Swollen River Series"

JOINTS

Joints are the connections between bones. They provide the space where one bone surface articulates with another for transfer of weight and energy. The shape of a joint affects its function. The function is affected by its shape.

All the joints in the body are involved in compression in weight bearing (decrease in the space between the bones), and elongation when being stretched (increase in the space between the bones). For example, there is compression between the femur and the tibia in standing; there is elongation between the same bones if you are hanging from a trapeze. There is compression in the shoulder joint in a handstand; elongation when you walk down the street. Joint receptors of the nervous system constantly feed proprioceptive information into the brain monitoring all the joints in the body. This information can be made conscious by focused attention on the joints. Do a body scan, and feel your joints in the positions they are in right now.

Joints participate in the dialogue between movement and stillness, falling and balance. The space between bones is the opening which allows mobility. There are three main structural classifications of joints: fibrous, cartilaginous, and synovial. In **fibrous** joints, there is no joint cavity and the bones are woven together with fibrous connective tissue. Some fibrous joints, such as the sutures of the skull, emphasize maximum stability and minimum mobility. Others, like the connections between the tibia and fibula of the lower legs, or the radius and ulna of the lower arms, are woven together with interosseous membrane and permit greater movement. In **cartilaginous** joints, there is no joint cavity and the bones are tightly connected by cartilage connective tissue with supporting fibrous tissue. In the joints between the bodies of the vertebrae of the spine, the connective tissue is a broad flat disc of fibrocartilage covered with fibrous tissues and four layers of ligaments. This allows considerable mobility of the spine while protecting the spinal cord and maintaining the vertical axis. The pubic symphysis is similarly connected with a fibrocartilage disc. In the cartilaginous joint between the epiphysis and diaphysis of the growing bone, the connection is immovable and disappears when bone growth is complete. **Synovial** joints include cartilage-covered bone ends articulating within a fluid-filled fibrous capsule **lined with synovial membrane**. Synovial fluid nourishes and lubricates the joint. Ligaments connect bone to bone and cover the joint capsule. Most of the joints we feel when we move our body are synovial joints: ball and socket joints of the shoulders and hips, hinge joints of the knees, ankles, elbows, fingers, toes, jaw; the gliding

Body Logic

I was visiting a friend who was a mechanical engineer. He had broken his ankle many weeks before, and it was slow in healing. I asked if he knew anything about the bones and ligaments which comprised the joint. He admitted that he hadn't asked the doctor, hadn't looked in a book, and didn't really want to know. He also assured me that the ankle was an outmoded joint, which was structurally inefficient and consequently dangerous. He did, however, want to be better fast. I began a detailed foot massage, traced the bones, felt the tissues carefully, and described the structure of the ankle and its relationship to the foot and the knee. I also massaged the soft tissues between the tibia and fibula, checked the axis from foot to hip, and observed his walking pattern. What became amusing and interesting to us both as we worked, was that he had never considered that there was a logic to his ankle, or that his movement patterns could affect his structure.

Arthrometric Model

Developed by
John M. Wilson, U. of Arizona

1. Central zone: Cartilaginous joints of the lumbar, thoracic, and cervical segments of the spine
   Function: Body positioning
   Motion in all directions is available, but the range of motion at any single joint site is limited. Motion of the vertebral column must therefore be cumulative, creating and promoting positional changes in the relationship of the torso to the pelvis and vice versa. This is axial motion, distinct from locomotor motion which the limbs are equipped to carry out with their synovial joints.

2. First perimeter: Synovial joints
   Ball and socket
   Function: Direction
   Note that only these joints, which are triaxial in biomechanical terms, are in position to direct the limbs into three-dimensional space as the positioning center has predisposed the movement toward fulfillment.

3. Second perimeter: Synovial joints
   Hinge (elbow is hinge and pivot)
   Function: Range of movement
   These biaxial, weight bearing joints cannot alter direction, but fulfill range of motion.

4. Third perimeter: Synovial joints
   Elipsoid, gliding/sliding
   Function: Movement refinement and shock absorption
   Again, biaxial joints, but small and capable only of refinement, not alteration of the direction into space already established.

5. Peripheral zone: Synovial joints
   Hinges, elipsoids, saddles
   Function: Articulation, manipulation, and indication

Note: The two "zones," central and peripheral, incorporate body segments that articulate by means of many cooperative joints: in column form in the central zone; arrayed in close-pack and radial form of the hands and feet in the peripheral zone. By contrast, the three perimeters intersect the articulating centers of particular joint sites.

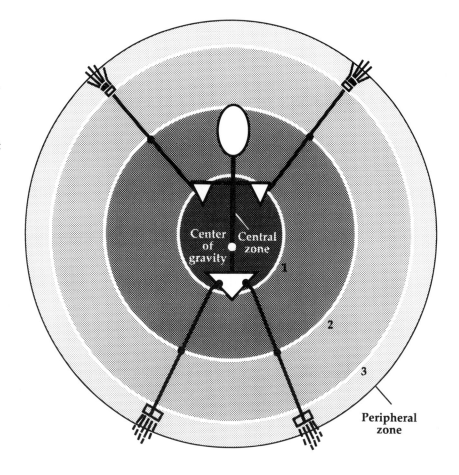

and sliding joints of carpals of the wrists and tarsals of the feet; pivot joints of the atlas on the axis, and the radius with the humerus; and the saddle joints of the thumb to first metacarpal, and big toe to first metatarsal.

If we use as a guide the Arthrometric Model described by John M. Wilson,* we see the relationship between structure and function. We begin in radial symmetry, with arms, legs, head and tail extended out from a center of the body corresponding to the center of gravity. In this position we can observe the patterning of joints and their sequential functioning in both movement efficiency and aesthetic expression. As Wilson says, "Whole body motion fulfilling both standards of efficiency and expressive beauty always originates at the center where the function of position (torso-pelvic relationship) is established, flows outward through the direction-giving joints of the hips and shoulders, through the elbows and knees, the ankles and wrists and at last the hands, feet and head."

We can bring awareness to the joints of the body by focusing on the bones themselves, or on the spaces between the bones. Our goal in alignment of the joints is even and equal distribution of weight and force throughout the body. ❖

* For further information on the Arthrometric Model, see John M. Wilson's, *A Natural Philosophy of Movement Styles for Theatre Performers* (Doctoral dissertation; the University of Wisconsin-Madison: 1973) pp. 122-129, or articles cited in bibliography.

*Arthrometric Model**
10 minutes

Standing: Focus on the **central zone** of the body – the cartilaginous joints of the spine. Stabilize your spine and move your periphery (arms, legs, hands, feet). Stabilize your periphery and move your spine. Feel how the positioning of the spine affects all other body movements.

○ Bring your focus to the ball and socket joints of the shoulders and hips, the first perimeter of the body in the Arthrometric Model. Use your hip joint to determine direction of movement. Move sideways, initiating at the hip; backwards, forwards. Notice how you open your leg, rotating at the hip joint to move efficiently to the side; how you open your shoulder to reach something behind you. Feel how the **direction** of movement is established by the ball and socket joints of shoulders and hips.

○ Focus on the hinge joints at the elbow and knee, the second perimeter. Position the spine, establish direction at the hip joint, and allow the **range** of the movement to come from the elbow or knee. Move forward with a tiny step; a huge step. Reach your arm towards someone eagerly; reach hesitantly.

○ Bring your awareness to the gliding joints of the wrists and ankles, the third perimeter. Feel the **movement refinement and shock absorption** which occurs through these small bones. Jump and land through the bones of the feet. Fall into a wall catching yourself with your hands.

○ Begin to move the hinge joints of your fingers, toes, and your jaw in the peripheral zone. Speak through movement with these joints of **articulation, manipulation, and indication**.

○ Run or move quickly in the space you are in, dodging objects or other people. Let your joints work for you. Experience the articulateness of the whole body.

* Term developed by John M. Wilson.

Drawing the joints
15 minutes

✎ Draw a simple stick figure, appendages radiating in an X from center. Identify the center of gravity. Indicate the joints of the body: spine, shoulders, hips, elbows, knees, ankles, wrists, fingers, toes, and jaw.

✎ Draw a circle around the central zone of the body, equidistant from the center of gravity. This central zone includes the cartilaginous joints of the spine and their function, carried out by the iliopsoas and abdominals, is to integrate and position the body for movement. All of the other joints we identify will be synovial joints.

✎ The circle of the first perimeter passes through the shoulder and hip sockets; these are ball and socket joints (with three-dimensional movement) and their function in the body is to establish direction of movement.

✎ The circle of the second perimeter passes through the elbows and knees; they are hinge joints (with two-dimensional articulation) for fulfilling the range of movement.

✎ The circle of the third perimeter passes through the gliding joints of wrists and ankles, allowing movement refinement and shock absorption.

✎ The circle of the peripheral zone passes through the small hinges of the fingers and toes, and their function is articulation, manipulation, and indication. Include the jaw in this zone.

○ Look at the drawing. Stand with your body in this position and image each zone radiating from center.

Moving from the joints
15 minutes

Lying in constructive rest: Begin articulating your skeleton with your awareness on the joints. Image the spaces between the bones, rather than the bones themselves. Begin with the toes; move from the hinge joints of the toes. Move from the gliding joints of the seven tarsal bones of the arch and ankles. Move from the space between the tibia and fibula of both legs. Follow any movement that comes. Do a dance of the hinge joints of the knees; feel their two-dimensional movement of flexion and extension around the curved surfaces of the balls in their sockets.

❍ Explore the full range of movement in the hip joints. Image the fluid lubricating the surfaces. Bring your awareness to the fibrous connections at the pubic symphisis and at the sacroiliac joints. Explore the collective action of the cartilaginous joints of the spine, with one disc between each vertebra. Feel the fluid within the spinal column. Move from the pivot joint between the atlas and axis; then from the rocking action of the occipital condyles of the skull with the atlas. Articulate the hinge joints of the jaw in their fluid-filled capsules. Remember the discs.

❍ Go to the connection between the manubrium and the clavicle, the joint which connects the shoulder girdle to the sternum and ribs. Move from this important joint. Explore the full range of the ball and socket joint of the shoulder. Follow any movement impulses you find there. Feel both the hinge and the pivot at the elbow. Do a dance of the elbow joints. Move the radius and ulna; connect this to the elipsoid and gliding joints of the wrist. Articulate the hinges of the fingers and the saddle joint of the thumb. Move from all the joints in the body.

❍ Come to standing. Bring your awareness to the bones, then to the spaces between the bones, then to the fluid which fills the spaces. Begin to jiggle the body by moving through the synovial fluid. Let everything move; remember your jaw and neck, the bones of your feet. "Hang out" by jiggling the synovial fluid within the joint capsules. Freeze the joints and try to move. Return to moving from your synovial fluid.

❍ To finish: Move from the bones; move from the spaces between the bones; then move from the joint fluid. Notice how the change in focus makes a change in movement.

Bima, a Javanese Wayang's (puppet) "shadow." A rigid body and six movable parts are worked by three rods, but manipulated with great vitality and nuance in elaborate ritualistic performances.

117

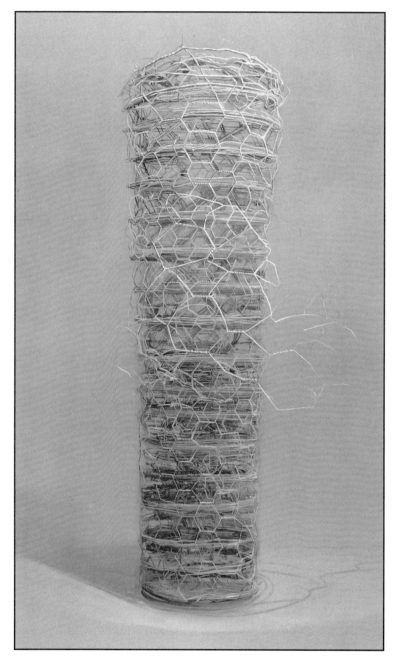

Painting: Jim Butler
Untitled

NERVOUS SYSTEM:
Body Listening

The nervous system governs the activities of the body. It works in conjuction with the endocrine system to direct and supervise body functioning. The functional unit of the nervous system is the nerve cell (neuron) characterized by its ability to generate and conduct electro-chemical energy forms called nerve impulses. The **central nervous system (CNS)** includes the nerve cells within the brain and spinal cord. The **peripheral nervous system (PNS)** is comprised of bundles of sensory and motor nerves which radiate from the brain and cord and reach to all parts of the body. Within the PNS, the nerves extending from the brain are called cranial nerves, and those from the cord are called spinal nerves.

There are two further distinctions within the nervous system which are particularly useful to our study, the **somatic** and **autonomic nervous systems**. The **somatic** nervous system receives, interprets, and responds to information related to both our inner functioning and our outer environment. The **autonomic** nervous system is responsible for our internal functioning. It affects the smooth muscle and glands of the vital organs (viscera) necessary for survival, including the lungs, the heart, and the digestive and reproductive organs.

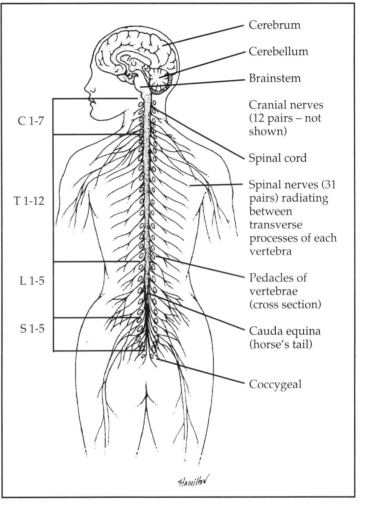

C 1-7

T 1-12

L 1-5

S 1-5

Cerebrum

Cerebellum

Brainstem

Cranial nerves (12 pairs – not shown)

Spinal cord

Spinal nerves (31 pairs) radiating between transverse processes of each vertebra

Pedacles of vertebrae (cross section)

Cauda equina (horse's tail)

Coccygeal

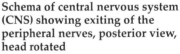

Schema of central nervous system (CNS) showing exiting of the peripheral nerves, posterior view, head rotated

The nerve cells for the autonomic system are considered part of the peripheral nervous system, and gather in ganglia (collections of cell bodies) in specific sites throughout the body. Concentrated groupings of these ganglia can be found in the area of the body along the front of the lumbar spine, interweaving with two muscles, the crus muscle of the diaphragm and the iliopsoas, so important for integrated movement. The autonomic (visceral) nervous system is often referred to as the "belly brain" in movement training techniques, and is considered to be the location for "centered" energy. The dialogue between the somatic nervous system – which is involved in our interaction with the outer world, and the autonomic nervous system – which is concerned with inner functioning, is fundamental to an integrated experience of the body – body listening.

Release

I have been hysterical only a few times. Once, I had shown a dance created for students at a choreographic competition. It was important to me, and I was eager to hear the response of the "judges" (adjudicators). They began by telling me that I had good dancers, but that they looked "emotionally repressed." They continued to tear the piece to shreds as my students sat around me listening. I began to cry – which was not the proper response. As I left, my students were angry at the attack on their work, but I was numb. When I got to the house of friends, I lay down on the floor and cried for many hours. My feelings seemed way out of proportion to the event, but I could not stop. I felt every criticism I had ever heard, and also felt my intense need for approval. As I cried, my body twitched uncontrollably. I was cold, and wrapped myself in a blanket. At the same time I had the strange feeling that I was observing the whole scene from a distance. My friends eventually got worried, and encouraged me to call a therapist who knew my work as an artist. The woman heard my story on the phone and said, "Stop. Stop crying." It was like a slap in the face. "You can come back to these feelings at another time, but right now, you should stop." And after a pause, "You've had enough. You might drown in your feelings." I had expected to be consoled. Instead, I stopped. I hung up the phone and washed my face. I learned that I could shift my mind. Later, I worked on the feelings involved in the experience.

The autonomic nervous system is also divided into two parts, the **sympathetic** and **parasympathetic nervous systems**. The **sympathetic** is activated when it is necessary to focus the body towards activity and survival, for example, in an accident, or during a race or performance. Stimulation of the sympathetic nervous system results in increased activity of the heart and lungs, and decreased activity of the digestive organs. Thus, when we are about to give a lecture, or run a race, our heart rate will speed up, while digestion is put on hold. The **parasympathetic** nervous system, on the other hand, is activated when the body is ready to relax and digest and there is time for integration, for example after a big meal. In this situation the heart rate and breathing slow down, and the digestive system is activated. A daily balance between activity and rest is fundamental to the healthy functioning of our internal organs. We all know the situation, however, of eating a big meal and then running off to a meeting or a physical work-out. In this case, the nervous systems send contradictory messages: The parasympathetic nervous system is activated to stimulate the digestive system for integration, and the sympathetic nervous system is activated to deal with high level functioning in the world. One physical result of these conflicting messages which is easily recognized is indigestion. If this imbalance is a constant occurence, the further result is nervous system exhaustion as the various components function inefficiently and are denied time for recuperation through rest.

Individuals who are under constant stress, or who have been "on guard" for prolonged periods (such as war veterans, students during exams, mothers tending young children, or children of alcoholic or abusive parents) often find it hard to regain a balance between the sympathetic and parasympathetic nervous systems. For many of us it is difficult to establish a situation which allows enough time, trust and relaxation to activate the impulse for rest, digestion and integration. Part of the process of healing the nervous system is to establish an environment where it is all right to let down and to be supported, without the pressures of alertness for survival. In some healing situations, such as working with an abused child, the individual may never have known a safe state, and the process is one of introducing the experience of relaxation to the nervous system. Within a safe context, the body will find its own natural rhythm between activity and rest. In bodywork the balance between introducing new information to the body, and allowing time for rest and integration is essential. Similarly, it is important to establish an environment that is conducive to a calm state. Sometimes if bodywork is quite deep, or threatening in any way to a sense of safety or stability, the autonomic nervous system will be activated for protection. This may result in increased blood flow to the vital organs and a sensation of cold in the periphery or in the whole body. It is useful to have a blanket available for additional warmth and to be aware of this important signal of body overload. It is better to reduce stimuli in this situation until the body has stabilized and is ready to continue; it is rarely useful to push through signals from the body which say "enough."

There is a dialogue between the **somatic nervous system** and the **autonomic nervous system**. As we receive information about our outer environment and make choices between activity and rest, we are affecting and affected by our inner environment. For example, if we are making an important decision, our stomach may become upset; if we are experiencing illness or indigestion, our involvement in activities may be affected; if we are in a warm, invigorating environment, we might feel energized. To a certain extent, we can affect our health by the choices we make moment by moment; our choices of environment, of work patterns and of relationships are among the many aspects which affect our well-being.

THE NERVOUS SYSTEM AND MOVEMENT

For movement to occur, there must be a signal, or impulse, from the motor cortex of the central nervous system to activate the muscle fibers. Although little is understood about the details of brain functioning, the nervous system gives the commands and the muscles work to carry out the tasks at hand. Reflex patterns, like lifting your hand off of a hot stove, occur through a reflex arc – the sensory nerves send a signal to the spinal cord (danger: this is hot), the interneurons connect the message through the spinal cord to the motor nerves, and an impulse is sent to the muscles of the hand and arm to respond (move your hand!). In a reflex arc the response is carried out at the level of the spinal cord, while the brain is being informed of what has occurred: you move before you are aware what has happened. In more complex motor patterning, such as riding a bicycle or typing, the learned movements are initiated and coordinated by various components of the brain. In a simplified view of motor activity, the decision to move is initiated in the cerebral cortex (I'm going to do a movement), the movement is begun and smoothed out by the basal ganglia (I'm doing the movement now), and then it is further refined and coordinated by the cerebellum (how efficiently can I do it?). These movement centers work in conjunction with other aspects of the brain. The cerebellum is also responsible for constant, instantaneous correction of every movement, and for maintaining muscle tone throughout the body, ensuring that muscles are able to respond when called to action. When we learn a new movement, or change an old pattern, like learning to use a new keyboard on a computer or learning to ice skate for the first time, the nervous system begins by using multiple, generalized motor pathways to affect a response. These pathways are gradually refined and the choices simplified for efficiency and speed. Thus learned movement goes through a phase of "awkwardness" until the nervous system has "sorted out" the most effective pathways. Once a pathway is open, the movement pattern remains in the body (sometimes called "motor memory"). For example, once you learn to ride a bicycle, the skill remains throughout your lifetime unless interrupted by illness or injury.

Because of the importance of the nervous system in movement initiation and patterning, we can work through the nervous system directly to affect muscle repatterning. There are several methods used in bodywork to affect neuromuscular efficiency. **Visualization techniques** can influence our picture of our environment and of our body, and thus affect our motor responses. For example, if you imagine yourself in a safe, warm place, with your head resting on a pillow, your neck muscles may

❖

In the years that followed, I watched several students go through a similar process. Each time was a pressured situation: A strenuous performance, a final exam, a decision about dropping out of school. The nervous system would be overloaded with expectancy and possible judgment, and it would pass its threshold for containment and begin to release "uncontrollably" in forms such as crying or shaking, laughing or talking, or extremes of heat and cold. Because I was familiar with this process in my own body, I was able to guide someone through the experience. I used firm touch to bring awareness and boundary to the body, allowed a certain amount of time for release – without panic or fear, and then would say, enough. This is enough for now. You can come back to these feelings at another time.

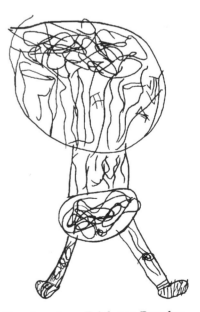

Drawing: Anya Brickman Raredon, age 6

❖

The image of a shaman, in my mind, has always been connected with the origins of performance. A shaman is a figure in many "primitive" cultures who is responsible for healing, for shaping rituals, and for guiding the community in ceremony and decision-making. His or her selection is often based upon unusual characteristics: a tendency towards isolation, to visions or epileptic fits, or to creative or healing powers. The basic idea is that the individual has been ill, moving beyond the realm of the normal, and survived, returning to share the vision with the community. Thus they are initiated as guides for healing. This is a journey of the nervous system: the capacity to vision, image and perceive realms beyond the normal senses, and integrate the experience with daily functioning. Not everyone needs to engage in the self-focused awareness common to the artist-healer. But in this view, the responsibility is to be a guide for others through the realms of the unknown.

relax in relationship to the image in the mind. If you are rehearsing for a dance performance or preparing for a ski race, you can visualize yourself doing the movement clearly and effectively; the mental picture will "rehearse" the neuromuscular impulses necessary for efficient movement. Conversely, a negative picture or self-image, or excessive corrections, can reinforce awkward patterns or "mistakes." For example, if you are told you "turn badly," and you begin to see yourself in that way, when you come to the moment to do a turn, the negative picture of yourself doing the movement may appear, and your nervous system will respond accordingly – you carry out the preformed image of "incorrect" movement. It is useful to have an integrated image of your body in motion; when corrections are given, take time to absorb the changes into the overall flow of a movement. When giving corrections to someone else, use language to describe an integrated body picture.

Images are useful to integrate multiple layers of an experience within a single exercise. For example, using the image of "melting your shoulders into the floor like butter" as you lie in constructive rest, can affect the release of neck and shoulder muscles and heighten your experience of gravity. Certain images work better for different individuals due to personal life experiences and habits, but some images have proven generally effective in the release of specific tension-prone areas of the body and in the development of movement efficiency. Lulu Sweigard, in her book, *Human Movement Potential, Its Ideokinetic Facilitation*, published in 1974, documents her research on the relationship of imagery to muscular release and repatterning. In contrast to the use of imagery, one can use **directives** to give a single command to the nervous system. Often, in relaxation techniques, awareness of muscle use is established by contrast of tension and release in various muscle groups. For example, a directive can be given (by yourself or by a group leader) to: tighten your jaw, hold the tension, then relax completely; to tighten the neck, hold the tension, then relax; to tighten the shoulders, then relax; tighten the forearms and so on. Especially when done lying on the floor, with eyes closed, in systematic progression through the body, this exercise can bring awareness of the sensation of tension and release in various muscle groups, and it can result in overall relaxation. Its effectiveness is related to using directives which affect muscle groups according to function. The use of directives can be frustrating and ineffective when directed towards the release of a single muscle in the body. Muscles respond to function rather than in isolation. The body is unable to respond to a command to "release the sternocleidomastoid muscle" of the neck; instead, either a functional directive such as relax your head to the side, or a visualization or image of relaxing your neck into a pillow, will affect the release of that muscle in particular, accompanied by all the muscle groups involved in head support.

Imagination exercises, such as guided journeys or experiences where you see yourself moving in an imaginary setting, are also useful in affecting the neuromuscular system in bodywork. An example of this work would be the following: Relaxing, lying on the floor with arms extended to the side and eyes closed, imagine the following: You are Sinbad the sailor and you are huge. Without moving your body, imagine you are reaching your hand into a long, narrow cave to steal a treasure. As your hand reaches further and further down the tunnel, it is extending and growing longer and longer. Your finger tips can almost touch the treasure. Your arm and hand stretches (in your imagination) a little bit longer until you can hold the treasure in your palm and bring it back. With no actual movement of the arm, imagine your hand returning, withdrawing out of the cave. Open your eyes; stand up and feel the length of your arm. Do the other side. Imagination can allow your fantasy to play and your nervous system to respond; it gives the opportunity for you to experience various body images and sensations without the confines of reality.

In my experience, there is a subtle difference in the use of visualization techniques, imagery, and imagination; and by extension different effects from the words "visualize," "image," or "imagine." Visualizing, to me, implies a visual distance – creating a picture separate from myself; imaging implies a visual picture, but I feel present in the experience; and imagining suggests that I am involved, but in a fantasy experience – that what I am doing is not really happening. Thus, in the context of the "To do" explorations in the text, I suggest that you "image yourself" doing an activity. Ideally, you can choose the language and experiences most effective for your own work. ❖

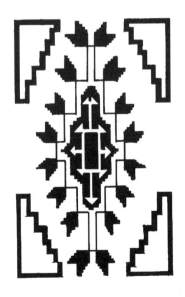

Center design of a Navajo rug

Observing the dialogue between the somatic and autonomic nervous systems
10 minutes each

In standing or seated position, image yourself in the following situations and observe your body responses:

❖ You are feeling sick and would really rather stay home and rest, but you are going out to a lively party. (or)
You really don't feel like talking to anyone, but you are in front of a class of students, giving a lecture.
The autonomic nervous system wants to be inwardly receptive, while the somatic nervous system is being outwardly expressive – conflict between rest and activity.

❖ You are in a class and have something very important to say, but the teacher never calls on you. (or)
You go home and want to tell your parents everything about your life, and no one is interested in listening.
The autonomic nervous system wants to be outwardly expressive (increased heart and breath, churning stomach), but the somatic nervous system is forced to be receptive, no expression – a scenario for ulcers!

❖ You have something to say, and you get to tell it to your best friend. (or)
You want to celebrate, and there is a great party going on.
The autonomic nervous system is outwardly expressive, and the somatic nervous system is outwardly expressive – ready to be active.

❖ You need to rest, and you choose to stay home and read a book. (or)
You are tired and you take a vacation by the sea.
The autonomic nervous system needs to be inwardly receptive, and the somatic nervous system is inwardly receptive – ready to take in.

We do versions of these experiences all day long. There is no right or wrong way of responding, but there is a dialogue between the needs of our body and the outer demands of the world. Observe your interactions throughout a day. How often do you respond to your own inner impulses, and how often do you make choices based on the needs or demands of others? Can you imagine a dialogue between the two?

Visualization, imagery, and imagination
20 minutes each

Lying in constructive rest, or relaxing in any position, eyes closed:

❖ Visualize yourself doing an activity efficiently and comfortably. For example, running through a park, walking calmly and comfortably through a room of people, having a clear conversation with someone you care about. Listen to the sounds around you; feel the responses occuring in your body; smell the scents in the air. Later, actually do the same activity. See if the visualization had any affect on your experience.

❖ Identify imagery that is useful to you to relax different parts of the body.
○ Identify negative images that you might have about your body.
○ Replace a negative image with a positive picture. Notice how this mental change affects your movement.

❖ Try an imaginative journey: Lying in constructive rest, or relaxing in any position, eyes closed:
○ Imagine yourself walking towards a door that leads to a stairway that goes down into the ground. Get a clear picture of the door and the area leading to it in your mind.
○ Go through the door and begin to walk down the stairs. Continue to walk down, further and further, until you arrive at a bottom. This may take a long time.
○ Imagine clearly the place where you have arrived. What or who is there with you? Notice sounds, sensations, smells. Spend some time and wander around or rest.
○ Find an object or an experience (talking to someone, seeing something) to bring back up with you.
○ When you are ready, leave the place you have found, and begin to climb the stairs. Come all the way back to the top, and go out the door. Begin to open your eyes.
○ Notice how your body feels in relation to the journey. Are you relaxed? Tired? Is there any response?

Photograph: John Huddleston
"New Haven"

DIGESTION

The digestive system is involved with taking in, integration, and letting go. It is one continuous tract from the mouth to the anus, bringing elements of the outer environment through the inner environment for digestion and absorption. The tissue itself is partially composed of contractile fibers to aid in the digestive process and to move the food through the tract. A series of sphincters permit passage and prohibit reversal of ingested food. Like breathing, digestion is an exchange with the environment; our inner-outer dialogue affects efficient functioning.

The digestive tract begins with the mouth: lips, tongue, and teeth. The throat is the first sphincter, simultaneously opening to the esophagus and closing the adjacent passageway to the oral cavity and the larynx/trachea. The esophagus is in front of the bodies of the seven cervical vertebrae and behind the trachea, forming the central core of the neck. (Swallow to feel this pathway.) The esophagus and trachea provide vertical support for the neck at the organ level. The esophagus passes between the lungs and behind the heart. "Heart burn" is a common term for indigestion occurring in the esophagus. It then passes through a hole in the breathing diaphragm to connect to the stomach. The diaphragm divides the torso into two cavities, upper and lower, and forms the floor for the heart and lungs and the ceiling for the stomach and liver. The stomach has considerable mobility, suspended by the lesser omentum, and is capable of extending as far down as the bladder after a full meal. The primary function of the stomach is to break down food for absorption. The pyloric sphincter oversees passage of churned food from the stomach to the small intestine. Ninety percent of the nutrients pass into the blood stream over a period of three to five hours while the food – now called chyme – is in the small intestine. The remaining ten per cent of the absorption process occurs in the stomach and large intestine. Small capillaries remove nutrients to the liver through the hypatic portal system. The small intestine, commonly known as our "guts," is approximately one inch in diameter and twenty-one feet in length, and consists of the duodenum, the jejunum, and the ileum. It intertwines in the cavity made by the frame of the large intestine, supported by the horizontal bowl of the pelvis and the front of the lumbar spine; fascia (common mesentery) contributes to their tight packing, and the abdominal sheath gives muscular support on the front surface and wraps around to the vertebrae in back. Vessels and nerves arise on the posterior abdominal wall and pass through the tissue to serve the organs. The large intestine continues from the small intestine: in the lower right corner of the pelvic bowl the residual appendix connects, the ascending colon travels up the right side of the pelvis giving

Gut Response

I was at a faculty meeting where we were discussing curriculum. Much reference was being made to the elimination of "gut courses." Dance and the arts in general are often included in this catagory. I spoke: "Since the gut is the place in the body responsible for digestion and integration, perhaps we should consider the essential value of these courses in a healthy curriculum. They require us to embody learning."

Passageway

Students in my anatomy classes wrote of their problems with anorexia or bulimia. One, who was recovering from anorexia, worked on her own to understand the digestive tract. First she visualized the continuous tube from mouth through anus; then she imaged relaxing the sphincters one by one. After a few weeks she could image the process of allowing outer material to pass through her body without experiencing fear of being hurt or need to grip for control. She said, "It helps to have some idea of where the choices are located. I have worked with this problem for two years and never had any idea of the shape of the digestive system. The inside of my body just felt dark and tight. Now I can imagine the passageway."

vertical support at an organ level, the transverse colon travels horizontally across our "belt" area, the descending colon travels down the left side, the sharp turn is the sigmoid colon, the rectum arches back and travels down the front of the sacrum, and the anal canal and anal sphincters are the final exit from the body. The sphincters provide areas of choice or gateways along the way. Muscular undulations along the tract serve, with gravity, to move food through the body. Problems can occur when there is tension, holding or indecision at any of the sphincters, or lack of tone in the smooth muscles of the tract.

As we learn to trace the digestive tract, we can relax or strengthen the sphincters and tissue through visualization techniques and exercises. Hands-on massage can also be used in many areas, particularly the small and large intestines. Bulimia, anorexia, ulcers, indigestion and hemorrhoids are a few of the illnesses concerned with the functioning of the digestive tract. The digestive system is one area where emotional problems manifest on a physical level and can be recognized. ❖

Schema of digestive organs and related structures

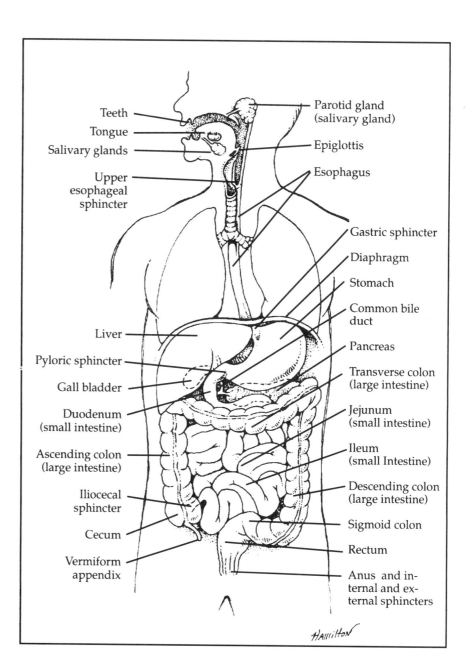

Tracing the digestive tract
15 minutes

Lying in constructive rest, with your hands on your belly:

❍ Bring your awareness to your mouth. Feel the jaw relax and the mouth fall slightly open. Swallow. Feel the passage of saliva down the esophagus, in front of the seven cervical vertebrae (behind the trachea). Image the descent through the diaphragm and into the stomach.

❍ Keeping your eyes closed, place your hands on the area below the ribs on your left side. This is your stomach. Gently massage this area to feel its density.

❍ Walk your hands across to the area below the right ribs. This is your liver; it fills the entire width and depth of your ribs on the right side.

❍ Bring your hands to the area below the sternum and xyphoid process often called the solar plexus. This is the area where your pancreas lies on a diagonal from the front to the back of your body. Breathe deeply. Feel the movement of the diaphragm as it presses down into the organs, massaging them with each breath.

❍ Return your hands to the stomach, on the left side of the body, below the ribs. Feel for a dense area near the angle of the ribs where the stomach meets the small intestine; this is the pyloric sphincter.

❍ Move your hands around the abdomen, feeling the inter-weaving of the small intestines. Your stomach may "gurgle" as you stimulate the digestive tract.

❍ Circle your hands clockwise around the perimeter of the pelvis; image the large intestine as it curves back towards the sacrum to become the rectum and anus.

❍ Contract and relax the anal sphincters, just as you contracted and relaxed the throat to swallow. Image the continuity of the digestive tract as one long tube from mouth to anus.

❍ Note if there are any areas particularly hard to visualize or feel in your digestive tract. Use your breathing from inside the body, and gentle massage from outside the body to help bring these areas into awareness.

Painting: Robert Ferris
"Two Swimmers"

HUMAN REFLEXES AND DEVELOPMENTAL PATTERNS

The first year of life is a continuation of the growth process begun in the womb. **Preverterbral patterns** are encoded in the body from our evolutionary heritage. **Neuromuscular reflexes** facilitate survival and underlie all movement patterning; they are gravity-related and include patterns for head righting, balance, and spatial reaching. The **human developmental patterns** guide the newborn through the journey from belly lying to standing. They provide a progressive sequence of movement for integration of the whole body. Perception of how long something takes, how far something is, and how much effort is needed to accomplish a task are established at the movement level and underlie our concepts of time, space, and energy: a baby crawls across a floor and picks up an object and learns about time, distance and appropriate energy use. Reflexes and developmental patterns progress in overlapping sequence. As one pattern is evolving, the neuromuscular coordination is preparing for the next. The senses motivate movement and provide the intention behind more complex patterning: smell, sound, taste, sight, touch. Early kinetic experiences are the basis for our understanding of self and other; communication is enabled through the common dimensions and capacities of the human structure.

PREVERTEBRAL PATTERNS

The developmental progression of the child parallels the evolutionary development of the species. Examples within this progression include the asymmetry of the single cell, radial symmetry of the starfish, bilateral symmetry of the shark, belly creeping of the salamander, four-footed crawling of the tiger, "hunkering" or sitting on haunches of the squirrel, "brachiation" or hand-opposite-foot tree swinging of the monkeys in preparation for our contralateral reach in walking. Prevertebrate patterns which are present at birth from our evolutionary heritage include: Breathing (internal/cellular and external/lung breathing), navel radiation (navel as center of nourishment and movement), mouthing (head as initiator of perception and movement with the body following), prespinal (undulating movement initiated from the digestive tract – mouth through the anus or from the brain and spinal cord within the skull and vertebral column). Similar to the transition from ocean to land, the bones and joints develop in strength, through the distribution of calcium, in response to body use. Bone compression in relation to gravity stimulates calcification centers. For example, as the baby moves from lying to sitting to crawling, the hip sockets gradually receive more weight and calcify in preparation for standing. When a baby is placed on its feet without time to measure its own way from floor to vertical or without time for sufficient skeletal and

Progress

I tried for many years to sponsor workshops in developmental patterns. Twice I scheduled guest teachers at our school; both times I left the classes before they were over. I had a dream: I woke terrified that I was being dropped from someone's arms. A colleague said, "If you don't bond with gravity as a baby, you never trust you are being supported. You have to hold on." I thought about support and the difficulty I had working with the developmental patterns. Because the patterns are preverbal, it can be hard to articulate confusion and allow change.

❖

I took a workshop with Caryn McHose. She had us begin by releasing our weight into the floor. She said, "You have to bond with gravity before you feel safe to push away. Once you release, you are free to stand."

❖

This year, I taught the patterns for the first time. I began by saying, "I have resisted this material for many years. It is not as easy as it seems. Lie on the floor, on your belly, and feel supported by the earth."

neuromuscular development, the results can include problems with perception, movement, or structural support accompanied by a sense of instability. Like taking an exam without having done the reading, you may know some of the answers and get by, but there is a feeling of insecurity about the information; something has been missed that will be needed later on. Reflexes and developmental patterns can be restimulated for adults, to provide a more solid foundation for complex movement patterns and emotional growth.

BIRTH

The importance of the head is established at birth. The body begins with the union of two single cells: the ovum and the sperm. Thus, the experience of cellular asymmetry rapidly progresses to the folding of tissue layers in the formation of the fetus. The baby develops for nine months in the womb, shaped in a C curve around the nourishing umbilical cord connected to the mother. The lifeflow comes through the cord to the belly, creating radial symmetry (like a starfish) with arms, legs, head and ancient tail (the coccyx) making six appendages from the center. Within the elastic membrane of the uterus, the developing fetus fits into the niches of organs and bones, pressing against the boundaries of the womb with its own body surfaces, exposing the familiar kicks and pushes visible and felt through the belly of the mother. The amniotic fluid gives constant tactile stimulation to the baby's skin and provides a gravity-free environment for the developing tissues. The membrane and wall of the uterus give periodic stimulation when the fetus is small, and constant tactile stimulation when it grows bigger. In most births, the fetus has assumed a head down position. This allows the head to press down upon the cervix of the uterus, the opening to the vagina or birth canal. The pressure of the head on the cervix and pelvic floor causes the pituitary gland to secrete more pitocin-crytocin which increases contractions. The physical structure of the baby's head provides something for the cervix to pull up around. The pelvic floor helps the baby's head to flex so that the smallest diameter of the head is lowest; it also creates a funnel for head rotation. Thus, the push and reach pattern of the baby's head stimulates the opening, or dilation of the cervix. As the head procedes down the birth canal, the body is propelled by the reflexive contractions of the uterus and the pushing of the mother's abdominal muscles. The baby passes between the muscles of the pelvic floor, and through the "hole" provided by the pelvis, and the head emerges into the world. The body follows, so the shoulders are aligned with the ovoid axis of the opening, rotating and spiraling as it slides out the opening of the vagina propelled by the strong contractions of the mother. Wartens jelly within the umbilical cord expands and stops the nourishing flow when exposed to air and change of temperature, making respiration through nose and mouth the new source of vital oxygen. The umbilical cord is cut, separating mother and child. The placenta, the lifegiving core of the womb, is expelled as the "after-birth." The birth process, known as "labor," is generally many hours; it can be as fast as a few hours or be prolonged over a period of days. This is one description of the birthing process; there are other possibilities including the experience of breech, feet-first, or cesarean section births. Separation from the mother begins our transition to bilateral symmetry,

Drawing: Nell Thorne, age 6

132

and the change from life in a fluid, gravity-free environment, to one with air, gravity, inertia, friction and momentum, and the world of physics as we experience it daily.

A majority of the proprioceptors providing information about body position in space are in the head and neck to protect and receive stimulation from this vital area. The movements of the head begin the first reverse curve of the spine, changing from the C curve in the womb to the eventual S curve of the vertical stance. As the cervical vertebrae make a forward curve to balance the weight of the skull over the ribs, the lumbar vertebrae also begin to realign. Eventually, through the process of lying, rolling, flying (lying on belly extending arms, legs, head, and tail off the floor, supporting weight on the ribs and belly, or more specifically, on the pancreas), sitting, and crawling, the lumbar vertebrae complete a forward curve to balance the neck in preparation for standing. This pairing of spinal curves: cervical and lumbar (neck and lower back), thoracic and sacral (ribs and pelvis), creates our responsive balance in verticality. One result of this pairing, as we have already seen, is that alignment of one spinal curve reflects the activity of its pair.

BONDING

Bonding is connecting. The newborn needs nutritional, emotional, and tactile nourishment and protection to survive the next stages of development; the emotional commitment of the parents to give this support and the baby to accept this support is called bonding. Initial bonding occurs by accepting the new air-filled environment through breathing and the parenting couple through nursing and touch. By extension, bonding includes other people, objects and experiences throughout life. Positive supporting, or bonding with gravity as part of the environment, underlies all movement reflexes and patterns. As we connect our center of gravity with the gravitational center of the earth, we establish the ability to push away and to move. A baby who for various reasons has difficulty bonding with gravity or with a parent, may develop a rigid or hypertoned body through resistance, or a flaccid hypotoned structure by disengagement, rather than gradated movement. Work with issues of bonding can happen at any time in life.

REFLEXES

Reflexes are simple patterns of movement that generally deal with flexion and extension of the body. The baby is cradled in the arms, continuing the C curve, but it is also placed horizontally on a bed or floor, or suspended in the air elongating the spine and body parts in relation to gravity. This begins the lifelong dialectic between flexion and extension of body parts; flexion is the folding of body parts towards center or the navel, and extension is the expanding or reaching away from center. Efficient movement gradates evenly between flexion and extension on all surfaces of the body. Neuromuscular reflexes for sucking and rooting (turning head towards or away from smells and tactile stimuli) underlie the movements for nourishment through the mouth, and neck-on-body or head righting (the impulse to keep the head vertical in any body position) underlies perception through head rotation and support. Some of the other reflexes which underlie our movement patterns at birth are: physiological flexion (drawing of whole body towards center in response to stimulation of the periphery – palms of hands or soles of feet) balanced by physiological extension (reaching whole body away from center in response to stimulation of the periphery); extensor thrust (extending limbs away from center in response to stimulation of the periphery), balanced by flexor withdrawal (drawing of limbs towards center in response to stimulation of the periphery), gallant reflex (small flexion on one side of body, letting the other side stay long), and the asymmetrical tonic neck reflex (lying on belly, eyes initiating, one side of the body extends in the direction of the focus, the other side flexes as the head rotates). Tickle the bottom of your foot, and observe flexor withdrawal; this is the basis for the "hip reflex" discussed in the chapter on the knee.

Reflexes are survival responses concerned with gravity, balance, and extension through space. Let's look at the progression of reflexes: 1) Positive supporting is the reflex after flexor withdrawal/extensor thrust, and is the foundation for all patterns of locomotion: as we connect our center to the center of the earth, we send energy down so that we can push away and go up – to lift our head, sit, stand, and walk. Negative supporting releases the push so we can move. 2) Righting reflexes also relate to gravity. They help us keep the head on vertical and maintain a continuous head and torso relationship such as the impulse to lift our head as we bend over or fall; the holding of the head on vertical as we roll on the ground. 3) Equilibrium reflexes deal with balance. In the dialectic between falling and not falling, there are reflexive choices whether to widen our base of support through protective hopping, protective stepping or protective extension of a body part to "catch our balance;" we can counterbalance our center of gravity with extension of a limb in the opposite direction of our fall, or we can let the center release towards gravity in navel yielding. 4) Spatial reflexes are more highly developed responses, and use the body parts as limbs to reach into space. For example, if we are falling, we can use the head to pull us through space rather than broaden our base or yield to the earth to stop the momentum. Then we are reaching rather than just falling or maintaining. At this level, every reach is a fall, is a stretch, is a balance, is a fall, is a reach. Extension comes as a response to yielding to the earth. As we have noted, one reflex underlies the next in overlapping progression. If we trust gravity and have positive supporting, and lose our balance and begin to fall, we can use a protective reflex; but without positive supporting as our base, without a trust of ourself in relationship to gravity, the reaction is towards holding or propping the body rather than towards movement: we might rigidify our limbs and lock in the joints as we fall, resulting in injury or breakage of a limb. As adults, we can also choose to override our reflexes. For example, the reflex might be to put our arms out in protective extension to catch our weight in a fall; we might choose to keep our arms against the body until the last moment for dramatic effect.

Early Mexican clay fertility figure of mother and child

134

In child development, one can tell that reflexes are integrated when: each reflex has an equal and opposite reflex; the underlying reflexes are present; the individual has moved on to more advanced reflexes. Reflexes are not lost as we become adults, they are integrated into our movement responses and they modulate each other throughout the body. In advanced degrees of movement, it is obvious to a trained eye when a reflex is missing; work can be done to restimulate reflexive patterns. An integrated reflex is no longer dependent on a stimulus; for example, we can do a passé in dance, or a deep hip flexion, without tickling our foot. Stimulation and perception are important aspects of early development. The baby explores through tactile information. Without the amniotic fluid as a source of stimulation, touch, heat and texture become important for establishing body awareness. Coordinations of eye to hand, the distance from the mouth to a hand or a foot, the focusing of the eye on parent or guardian, the crossing over of the thumb in juxtaposition to the fingers for grasping, are a few of the many developments which occur as you watch a newborn explore their new environment and establish their sense of self and other. Neuromuscular coordination and strength increase in response to motivation. Stimulation and perception motivate movement.

HUMAN DEVELOPMENTAL PATTERNS

Human developmental patterns are about integrating the whole body. Built on the reflexes, they begin by connecting head to tail; then they integrate the upper body, the lower body; then all the parts on the same side of the body (eye, arm, torso, leg); the opposite side; and finally the whole body by cross patterns through the center, in what is known as contralateral movement. Walking, skipping and spiraling are integrative movements based on contralateral patterning. Perception and intention motivate developmental movement. Sight, smell, taste, sound, touch stimulate movements towards and away from, and the impulses to reach, to pull, and to push away. The need to establish boundaries between self and other; and to locomote, to interact with our environment, and to communicate literally move us through the progression to standing. The developmental patterns alternate between bringing us into center, and taking us out into space, through flexion and extension. Connection with gravity towards weight and rest and reaching into space towards lightness and activity provide a span of dynamic range and health in the body. Six push patterns, and six reach and pull patterns, as identified by Bonnie Bainbridge Cohen* prepare the body for vertical support and mobility. **Spinal push** from head to tail/tail to head, **spinal reach and pull** from head to tail/tail to head; **homologous push** from two hands to two feet / from two feet to two hands, **homologous reach and pull** from two hands, two feet; **homolateral push** from one hand to foot of the same side/from one foot to hand of the same side; **contralateral reach and pull** from one hand through to the opposite foot/from one foot through to the opposite hand. The transition to standing is a combination of all the patterns, initiated with a reach of the head and hands. Balance is a dynamic relationship between gravity pulling us down to the center of the earth, and antigravity pushing us away from the earth. Centripetal force pulls us towards the ground and centrifugal force spins us out into the universe. Both balance and locomotion involve a constant process of fall and recovery, supported by neuromuscular reflexes. Rest and activity are necessary for the health of all our body systems. ❖

* For further information on reflexes and developmental patterns, see Bonnie Bainbridge Cohen's articles in the *Contact Quarterly*, Vol. 14:2 and Vol. 14:3, "The Alphabet of Movement."

Moving through the developmental pattern sequence
30 minutes

○ **Head to tail push pattern:** In the deep fold, C curve position on hands and knees (use padding under knees if necessary), forehead on floor: Initiate the push from the tail and spine to roll forward to the top of the head, sequentially through the vertebrae to an arched curve, like a bridge; push against the floor with the head to move back through spine to tail into deep fold; use relaxed, folded arms and legs as support (as "pontoons") but not as initiation. Repeat several times; sequence through each part of the spine. Feel the top and back of the skull on the forward curve, as though you are looking at your belly button. Feel the push of the skull to initiate the backwards sequencing to the tail. Image the baby in the womb, pushing with the top of the head.

○ **Head to tail reach and pull pattern:** Lying on belly or back, limbs extended: Reach and pull with the head and spine, reach and pull from tail; this creates a rocking action from head to tail, tail to head. Include motion from the heel or toes to increase the stimulation.

○ **Homologous push from the upper body:** Lying on belly, legs extended, hands by shoulders: Keeping head alert on the spine, push down simultaneously from both hands and arms until torso and head are lifted off the ground; legs stay relaxed. Feel the modulation through the joints of the arms and the support of the chest. Let this same movement propel you backwards in space.

○ **Homologous push from the lower body:** On belly: Find inchworm movement motivated from pelvis and coccyx for forward thrust through the spine and head.

○ **Spiral Roll**: On belly: Roll onto back initiating with head. Return to belly, also initiating with head.

○ **Homologous reach and pull from the upper body:** In squat position: Propel yourself forward in space by equal reach of both hands; land on hands (leap frog motivated from arms, chest).

Head to tail

❍ **Homologous reach and pull from the lower body:** From squat position, weight on arms: Thrust backwards in space propelled by simultaneous kick from both feet and tail, land on feet (leap frog movement backwards in space; traveling donkey kick).

❍ **Homolateral push from the upper body:** Lying on belly: Push backwards in space through the hands and arms loading weight towards, emphasizing, one body side at a time.

❍ **Homolateral push from the lower body:** On belly: Push from one foot, extending through the body and out the arm on the same side. Alternate sides. (salamander)

❍ **Contralateral reach and pull from the upper body:**
Belly on the floor: Move to hands and knees by crawling forward initiating with one hand, opposite knee coming forward for support. Alternate sides.

❍ **Contralateral reach and pull from the lower body:**
Crawling backwards: Reach with one foot, opposite arm comes back for support.

❍ **To stand:** From squat position: Reach with the head, reach with the tail; alternate energy pulls until you come to vertical.

❍ **To walk:** From vertical alignment, feel postural sway; initiate movement by reaching with one hand. Feel homolateral (same-sided) walking. Example: right foot and right hand swing forward simultaneously. Feel contralateral (opposite-sided) walking. Example: right hand and left foot swing forward simultaneously with rotation at the waist.

❍ Identify which patterns are easiest for you, which seem awkward or unclear. Spend time moving through the progression; work with individual patterns. Be patient. Work with developmental patterns returns you to a "baby" state of mind. Because it is preverbal, it can seem irrationally frustrating; it can also be a source of freshness and fun.

❍ Write, draw, talk to explore the work further. Be sure you feel supported and safe in your environment.

Homolateral push from lower body

Exploring underlying patterns
30 minutes

Lying belly down on the floor in an X: Release your entire body surface into the earth. Image your center of gravity. Connect it to the center of the earth. Experience **bonding with gravity.**

○ With your awareness on your center of gravity, feel **navel radiation**. Roll over, leading with a hand or a foot, in a cross pattern, returning to the X.

○ Continue rolling, leading with the head and eyes; experience **neck-on-body reflex** and **head righting**. Play with various movements initiating with the head. Initiate with the body and observe the response of the head. Surprise yourself.

○ Bring your awareness to the mouth. Work eyes closed at first. Feel the impulses for sucking by opening and closing your jaw. Follow the movement of the mouth. Feel your lips. Let the body respond.

○ Explore mouthing by feeling the impulse for sucking. Gently brush your lips against the floor for stimulation. Open and close the jaw. Follow the movement of the mouth; let the body respond. Be aware of smell and taste as you work.

○ Begin **pushing**; use any appendage: head, hand, foot, tail. Explore pushing.

○ Change your motivation to **reaching**; reach with each body part. See how this affects your movement.

○ Let your reaching lead to **stretching**, lead to **balance**, lead to **falling**. Let this move you through the room.

○ Explore falling. Begin with navel yielding, allowing your center to yield to the floor; soften your joints and roll out of the fall to stand. Fall again, by yielding your center; then roll, stand.

○ Walking or running slowly through space, alternate ways of falling with your locomotion. Explore the dialectic between up and down. Try using body parts as limbs to extend you into space as you fall.

○ Explore **spirals.** See how they affect your movement.

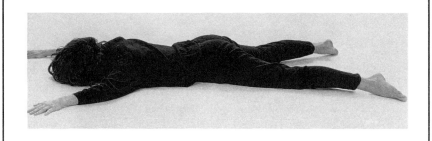

Bonding with gravity

Modulating opposites

5 minutes each

○ **Flexion and Extension.** Begin in a C curve lying on your side; Fold all the body parts towards center: head, toes, legs, arms, spine. Initiating with your feet, move into extension through all the joints sequentially. Roll to the other side and repeat flexion. Beginning with the feet, sequence through extension and back to flexion. Be aware of the modulations between full flexion and full extension.

○ **Down and Up.** Begin lying on the floor eyes closed: Come to standing. Take as long as you need. When you are standing, begin yielding to the earth. Again, take as long as you need. Add motivations for coming to standing, for going to the floor. Be aware of the modulations between release into gravity and standing in vertical.

○ **Reaching and Pushing.** In a comfortable position: Begin by reaching towards something or someone; when you accomplish your intention, begin pushing away. Explore the modulations between reaching and pushing.

○ **Opening and closing.** In a comfortable position: Begin to close the body; bring the body to a state of openness; return slowly to a closed position; explore. Feel the modulation between opening and closing.

○ **Together and apart.** Moving in space: Come together with someone or something; leave and go apart. Be aware that you can be apart when you are together; you can be together when you are across the room. Explore this dialectic of coming and going, meeting and leaving with various objects or people. Feel the modulation between being together and being apart.

○ Write, talk, draw to articulate your experience.

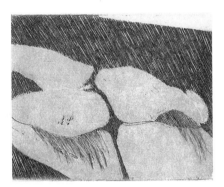

Etching: Harriet Brickman
"Iphimedeia Series"

SEXUALITY

Sexuality invites union with self and with another. **Sensuality,** as we have seen, involves the sense organs for sight, sound, smell, taste, and touch activated through the nervous system. These senses, sometimes referred to as the five "sense doors," inform sexuality by providing awareness of the constantly changing relationship of the body to itself and to its environment. **Sexuality** involves the reproductive organs of the body. The functioning of these organs is intertwined with sensual stimulation and is activated by release of hormones from the endocrine system. The potential for reproduction and for nurturing is inherent in our physical being and affects our experience of self and our interaction with others.

Respect for our physical selves supports sexual responsiveness. The more comfort and trust we have in our bodies, the more choice we have about our sexual actions and interactions. As we explore the territory between the "yes" or "no" of opening and closing, we can establish healthy boundaries for communication and expression. Respect for others is based largely on this respect of self. Perhaps nowhere more than in intimate communication with another do we need to know our bodies are with us, supporting our needs, and rejecting what is not needed or appropriate.

Sexuality returns us to childhood. Because responses to erotic stimulation are based in infant experiences of oral, anal, and tactile stimulation, they connect to powerful emotional needs which may or may not have been met as children. Sensual delight is generally alive and well at birth, and gradually becomes socialized within what hopefully are healthy restraints and boundaries. Thus, sexuality remains charged with issues of childhood as well as with an emerging sense of self and other as adults. Feelings about dependence and independence, abandonment and the need for control are a few of the possible concerns which affect sexual behavior.

The modulation of flow between being closed and being open is the pulsing life of healthy sexuality. There is no reason to be open all of the time; or to remain closed. The body receives subtle information from its environment and may know more than we can consciously understand about what is safe and what is threatening. As we trust the wisdom of the body to close when it is appropriate and to open when it is ready, we develop a different understanding of experiences such as "frigidity" or "impotence" as well as of "orgasm." To love, we need our whole self present. This adaptability gives us our responsiveness, and allows us to trust that our bodies will and can protect us if we listen to the signals.

Voyage

In my years of studying dance, I was always told that ballet is the epitome of femininity. The youthful sylph; the spiritual essence. In my own experience, ballet was the closest I could get to experiencing myself as a male in the world. The rigorous training, the concentration and competitive drive, the physicality and sweat gave me independence and distance from traditional girl responsibilities. I watched my father, as a watercolor painter, at his desk, focused for hours on his work. I went to ballet class daily after school and was required to focus equally as hard; school never provided that challenge. I was away from my small community, in the city, in the world alone, dealing with the challenges of mastering the body. And the occasional experience when the music and the dancing flowed together moving the body by forces I never knew existed, felt like religion to me. Obviously, it was an escape as well as a goal. My image was not to be a dancer, not to be feminine, it was to be free.

Sexual characteristics differ between female and male. The sex determination of the embryo occurs genetically. The primary reproductive organs begin in the developing embryo as the same structures, the gonads. By about the eighth week, they clearly differentiate into the ovaries in the female and the testes in the male. The ovaries remain inside the abdominal cavity, suspended by ligaments. The testes descend into the scrotum, outside of the body cavity, just prior to birth and are stimulated by the male hormone testosterone. The primary structures of the female reproductive system include two ovaries, the uterus, the Fallopian tubes, the vagina, the external structures of the labia, clitoris, and mons pubis, and the mammary glands (breasts). The primary organs of the male reproductive system are the testes within a sac of skin called the scrotum, the penis and other supporting structures, plus ducts and glands such as the vas deferens and the prostate glands. Principle endocrine secretions affecting sexual characteristics and functioning include the hormones estrogen and progesterone secreted by the ovaries, and testosterone, secreted by the testes. Comparing the female and male characteristics, the female has breasts, a higher percentage of body fat, a monthly ovulation cycle or "period," a wider pelvis, and a higher voice pitch. The male has a

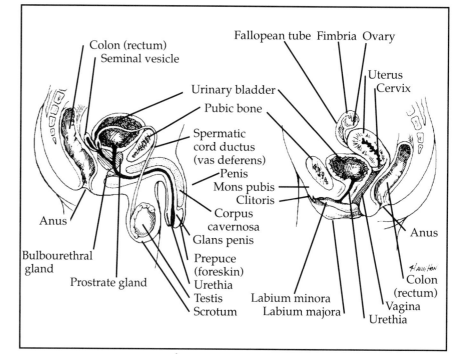

**Reproductive organs:
male and female, lateral views**

larger percentage of muscle than fat, a heavier bone structure, increased body hair, wider shoulders and narrower hips, and an enlarged thyroid cartilage of the larynx producing a deeper voice. The physical characteristics are further defined by cultural patterns and traditional views of male and female activities, which affect body image and physical development.

Sexual union in the human species involves personal interaction. In relation to our evolutionary past, face to face, front surface to front surface copulation enhances intimacy and the need for communication. As one of the many possibilities for making love, eye to eye, belly to belly connection brings us into a direct and vulnerable relationship with our partner. Dimensional thought invites reflection on past, present and future experiences, and acknowledges concerns and options about birth control. As sexual union expands its role from reproduction to pleasure and re-creation, it becomes a vehicle for physical dialogue with the potential of spiritual union (whole self to whole self).* Because of its connection to the endocrine system, sexuality is highly interactive with our emotional lives. Sexual experience can include the many subtle aspects of sexual exchange which we experience each day, based on energies awakened in ourselves as sexual beings. For example, the warmth felt seeing a friend laugh, the feelings stirred hearing a poem read aloud, or the empathy felt watching an athlete or a memorable performance, can involve our sexual energies. As we embrace our potential as sexual beings, we can begin to recognize the wide range of possibilities for stimulation and response which we encounter every day. Our sexuality is with us for our lifetime, affected by our choices concerning childbearing and parenting and living a dimensional life. Thus we can enlarge our view of sexuality to include our full sensual and sexual selves.

Although it is not within the parameters of this text to attempt to define or illuminate the nature of "sexual energy," it is helpful I think to acknowledge its presence. As we contemplate the effects of energy on the body, we can enjoy the choices we have for channeling this energy towards useful ends. Often, the tendency to flip from being "in control" to being "out of control" of our bodies, is based on a lack of awareness of the energy spectrum within each human being. The energies which arise within a body are of the same root, and we are the ones who choose their direction for release or transformation. The focused energies used to write this book, win an olympic race, create a work of art, rise to the top of the corporate ladder or negotiate a peace treaty come from the same source as the drive to beat a child, initiate a war, or hurt a friend. One of the challenges of our species is to be responsible for the use of our energy. And to be responsible, we must know our potentials. The highs possible within the natural makeup of the body are extraordinary and available to every person. So are the depths. Whether our energy source is labeled "sexual energy," or "emotional energy," or "competitive drive," or "love," we can become familiar with its characteristics and manifestations, and increase our capacity for ethical use. This awareness is different than the desire to "control" the body and the resultant need to get "out of control" by abusive use of sexuality, as well as stimulants and depressants. Instead of control, we seek a dialogue with our bodies, and a healthy respect for the vastness of our own nature. Then we can learn from ourselves. ❖

* Some cultures and spiritual practices developed techniques to channel sexual energies towards heightened awareness, connecting sensuality and spirituality. Studies in Kundalini yoga, shamanism, tantric buddhism, meditation, sufi spinning, as well as drumming, chanting and dancing are but a few of the many transformational traditions which explore the possibilities inherent in our nature.

Chinese talisman to vitalize the kidneys (symbolizing the female sex organs) – shapes suggest a pair of female dancers with "yin" receptacles. Attributed to Ling-pao, 12th Century.

143

Relating to self
15 minutes each

❖ Lying in constructive rest: Do a body scan. Claim your whole body, part by part. Continue with body painting and a proprioceptive warm-up. Stimulate each body part through touch and movement.

❖ Lying in constructive rest: Lift the pelvis to a diagonal position between your knees and your shoulders, like a giant slide. Allow the organs to release towards your thorax in relation to gravity. Lower the spine sequentially, yielding each vertebra to the ground.

❖ Image the pelvis as divided into four quadrants like a pie. Lift it one inch off the floor and circle it slowly to the lower right quadrant, then circle to the top left quadrant, then to the upper right, then to the lower left making a figure 8. Place the pelvis on the floor. Reverse.

❖ Begin small undulations from the tail to the head. Feel the energy moving up and down the spine. Roll to your side and come to seated position, legs crossed. Feel your breath. Image the energy channel along the front surface of your spine, from tail to the top of your head. Walk up the front of your spine in your mind's eye, opening any doors to places which feel closed.

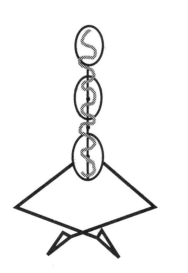

Plumb line with energy up the spine

Relating to others
15 minutes each

❖ Locate the center of your body; locate the periphery. Feel the dialogue between your core and your ability to reach out through the body to others. Establish your ability to say "no" to energies which feel invasive. Find a movement or a position or speak the word to clearly establish this boundary.

❖ Explore your ability to say "yes" and open your boundary to energies which feel good, useful. Do both simultaneously, taking what feels good, while filtering what feels negative.

❖ Practice being direct with your needs. Use language to help. Speak exactly what you are feeling right now. For example, "The noise drives me crazy." Then insert the words "in me" in the core of your sentence. "The noise in me drives me crazy." See how this affects your ability to respond to the situation. What do you do when there is too much noise inside yourself? What are your choices?

❖ Practice claiming your own experience. Again, use language to help. In Authentic Movement, after we watch someone move, the watcher speaks in response to the comments of the mover. Once the mover has spoken, the watcher says, "In my experience, I felt sad when you touched your head. In my experience, I was afraid when you stomped your foot." This is a way of claiming that every experience is your own perception. You can't know what another person feels. You can only know what you feel, stimulated by the other person. It would be very different to tell the mover: "You were angry when you stomped your foot." You are taking the experience away from the person who had it. The responsibility is to claim your own feelings. Then the person has a choice how to respond to your comment.

Raked sand and rocks in the South Garden of the Hojo, Tofuki-ji Temple, Kyoto, Japan

Painting: Robert Ferris
"Tom Bailey"

EMOTIONS

Emotions motivate and integrate the body. Often associated with "gut responses," emotions are part of our daily body functioning. The limbic system of the brain is involved with the emotional aspects of behavior related to survival, including pleasure, pain, anger, rage, fear, sorrow, passivity and aggression, and sexual feelings.* Experientially, we may feel such signals as heat flashes or sweating palms, tense muscles or cold feet. Unexpressed emotions are stored in the body. Each of us has places where our emotional tensions are held: the shoulders, the stomach, the lower back, and the jaw are common areas. As we move the body, we can release stored emotions, opening the mind to new possibilities. As we give expression to emotions, we can initiate the release of tension in the body, expanding our potential for health and movement. We have choice about emotion. Within the cerebral cortex, along with the sensory and motor areas, are "association areas" which are concerned with emotional and intellectual processes including memory, emotions, will, reasoning, judgment, personality traits, and intelligence. In the timing between sensory stimulation and reaction, the association areas inform our response. Emotions are often given extreme consideration: exaggerated in importance or overlooked altogether. Because emotions manifest as physical sensations in the body, we can use experiential work as a guide to emotional awareness. As we feel sensations register in our body, we can decide our response. Emotions are part of our human experience; without dominating our choices they can enliven our lives.

For some individuals, emotional feeling and expression is the clearest form of communication. Some of my friends distrust their bodies, but they are absolutely clear in their emotional responses. They trust their feelings over any other form of judgment. Others intuit what is happening long before they can pin it down in words or feelings. Some of us perceive physically. I learn most efficiently by translating all experience through the body. Thoughts and emotions register as movement. And I know other friends who would prefer to read about something first, get everything organized, and then enter the experience. My sister tells me she doesn't believe she has experienced something until she has read it in a book. Growth comes as we expand our possibilities for awareness. We are all different, and yet each person has the potential for the entire range of human experience. ❖

Support

In college, I made a painting of a geranium. It was late at night, and no one else was in the studio. I got transported amidst the twisting angular stems, the brightness of the blossoms, the darkness of the window behind. Hours passed. The next morning I looked at my work; it was the first time I had used emotion to motivate the motion of painting. It was alive.

❖

A young woman in our dance company was in a car hit by a drunk driver. We were told she would die. She was a very beautiful dancer, and extremely bright. When I first heard by telephone, I said, "No!" Later in the emergency waiting room, I saw the man who had hit her, arms flying and words pouring out about how someone had gotten in his way; how they were going to get it. I went up directly to his face, furious, and told him to stop lying; he'd perhaps just killed someone. And then I cried and cried, throughout the night as the various surgeries were performed. It was my first experience with moaning, calling out, bonding with others in sorrow and hope. This was the classic progression I am told: denial, anger, then feeling. She lived, and can now walk again after eight years of work. Brain damage affects memory and vision, so she can't read, can't dance. She says she does not want to be alive. She is very angry; the driver was fined $25.00 and released.

❖

I was in New Zealand. I stomped for miles along a beach before I realized that my feet ached so much I could barely stand. I felt hurt. It took physical pain to recognize I was in emotional pain.

---

*Gerard Tortora and Nicholas Anagnostakos, *Principles of Anatomy and Physiology*, pp. 319-320.

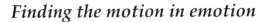

Finding the motion in emotion

○ Standing, eyes open: Begin by running or moving quickly in the room. Raise your heart rate; feel your blood pulsing. Continue your movement and include push-ups, aerobics, jumping-jacks, contact with the floors or the walls or other people, wiping your sweat, playing energetically; be direct with your movement. Make sounds if you want; call out to someone, laugh, talk, sing. Feel the easy release of emotion through motion.

Image work: masks
30 minutes

Relaxing in constructive rest, eyes closed:

❖ Imagine that you are putting on a mask of beauty. Let the mask be any image which represents beauty to you. Feel how beauty affects your whole body as you continue to lie in constructive rest. Breathe deeply. Remove the mask.

❖ Imagine that you are putting on a mask of rage. Create a mask that represents rage to you. Feel rage affecting your whole body in your imagination. Breathe deeply. Remove the mask.

❖ Imagine that you are putting on a mask of fear. Make a mask for yourself which represents fear. Feel the mask affect your whole body as you breathe deeply. Remove the mask.

❖ Imagine that you are putting on a mask of joy. Design a mask for yourself that represents joy. Observe your body as it experiences joy. Remove the mask.

❖ Imagine that you are putting on a mask that is funny, that represents humor. Let yourself experience what it feels like to be funny. Breathe deeply. Remove the mask.

❖ Imagine that you are putting on a mask of sadness. Create a mask for yourself that is very sad. Feel the sadness in your body. Breathe deeply. Remove the mask.

❖ Imagine that you are putting on a mask of ugliness. Create a mask for yourself that represents ugliness. Feel ugliness in your body. Breathe deeply. Remove the mask.

❖ Imagine a mask of any emotion of your choice. Create the mask very specifically. Feel the emotion in your body. Remove the mask.

❖ Feel the mask that is your own face. Imagine following the features of your face to paint a mask that enhances who you are. Breathe deeply. Remove the mask.

❖ Imagine a mask growing out of your face that is an exaggeration of what you are feeling right now. Feel that exaggeration in your whole body. Breathe deeply. Remove the mask.

❖ Finish. Roll to your side, come to seated. Draw, write, or talk about your experience. Consider that all the masks are different aspects of yourself.

❖ Repeat this same exercise, but allow your body to move. Image each mask, one at a time, and follow the movement of your body.

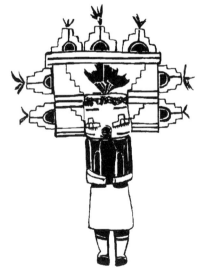

A Kachina doll (Cloud) representing one of 350 supernatural beings, emphasizes the masks worn by Hopi dancers.

Moving contrasts
10 minutes each

Beginning in any position which is comfortable to you:

❖ Move from the contrast between beauty and depression.

❖ Move from the contrast between fear and rage.

❖ Move from the contrast between joy and humor.

Be aware of the physical sensations caused by each experience. If you are working in a group, repeat; split into two halves and watch each other work. Change the contrasting pairings; for example, combine depression and joy.

Creative response
20 minutes each

❖ Select two pictures by thumbing through a painting or photography book. Look for images that reflect or evoke differing emotions. Write or move from your response to the images. Allow the contrast between the pictures to inform your expression. Say or move whatever comes to you. If you feel ready, read or show your work to someone else.

❖ Select two passages from writings that reflect contrasting emotions. Read them aloud. Move, write or draw whatever emerges for you from hearing the words.

❖ Select two newspaper articles or television broadcasts which reflect the world around you. Use the stimuli as motivation for writing, moving, or drawing your emotional response.

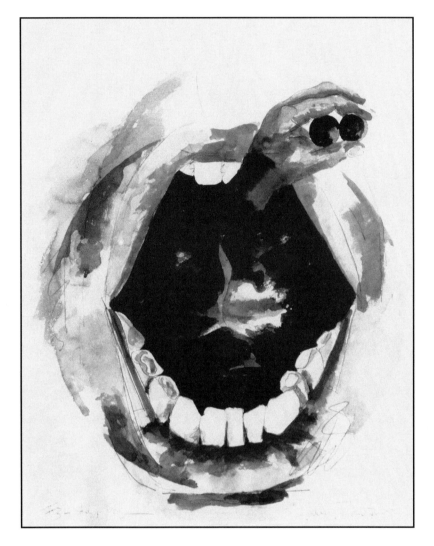

Painting: Gordon Thorne
"Study for Mary"

SOUND AND MOVEMENT

Sound gives expression to the body. The vocal folds (mucosa lined ligaments or "cords") run parallel to each other, from the front to the back in the larynx (from the thyroid cartilages, to the arytenoid cartilages). Thus, they are horizontal to the ground. We can touch this area in the front of the neck, and feel the vibrations as we make sound. The vocal cords are abducted (pulled apart), or adducted (drawn together) by movement of the arytenoid cartilages. In breathing, the vocal folds are abducted allowing the easeful passage of air; in phonation, or sounding, they are pulled taught (adducted), creating a thin space for the air to pass through while vibrating the cords. To add sound to the process of breathing, allow the air to pass from the lungs to the trachea, vibrate adducted cords in the larynx, and hear and feel the sound emerge. When you have a cold, excess mucus irritates the cords, thickening them so they vibrate slower. This creates the familiar low voice. The muscles of the neck, throat, and belly need to be relaxed to allow efficient use of the vocal mechanisms.

Three components affect pitch: the size of the space between the cords, the tension of the cords, and the amount of air pressure as it passes through the cords. The arytenoid cartilages reflexively adjust both the size of the space between the cords and the tension of the cords as we intend to sing or speak a particular pitch. Too much tension in the muscles of the neck and throat can prevent this reflexive adjustment from occurring efficiently. The diaphragm controls the amount of air pressure as it passes through the cords. Maintaining an expansive sense of relaxed tone in the belly and back helps the diaphragm create just the right amount of air pressure needed for the sound we intend.

Volume and timbre of sound are created by the various resonators in the body. Different body tissues have different resonant characteristics. In general, high pitches need small resonating chambers to amplify them because they have short wavelengths, and low pitches need larger resonating chambers because they have longer wavelengths. With that in mind, you may find it easier to resonate high pitches in your sinuses and lower pitches in your belly. Human sounds are composed of fundamental (main) pitch and overtones (higher "partial" pitches). Overtones give the pitch its timbre or tonal color. For instance, speak "ee." Then speak "oh." Notice that you feel more resonance in your sinuses as you say "ee" and more resonance in your belly when you say "oh." At an early age we learn to amplify higher overtones in a sound to produce "ee," and lower overtones to produce "oh." We change from one vowel to another by changing the timbre of the tone.

Silence

My way of perceiving the world was silent. As children, we would pile out of the car on vacations, line up on a rock and stare scenically and silently into the distance. My father was a painter, my mother a first grade teacher. Somehow sight dominated sound in my early experiences. We have many such photos in our family album.

❖

When I was teaching in Vermont, I would get raging sore throats two or three times a year which would turn into coughs, colds and go away after a few weeks of misery. Although I feel articulate in movement and writing, speaking feelings directly was difficult for me; the strain registered in my throat. During one such bout, I was walking in a field filled with crows. I was frustrated, and their cawing was relentless. I began cawing back, imitating their sound and enjoying the feeling it created in my body. My sore throat went away. After this, when I started my sore throat routine, I would, timidly at first, reply with my crow caws. Perhaps the vibrations loosen the tissues, or the unexpressed feelings get released by the sound; whatever the cause, my sore throats have gone.

We increase or decrease the volume of a tone we sing or speak by adding or subtracting resonators. Take a normal breath and say "Hi!" lightly in your nasal sinuses. Say "Hi!" again allowing the sound to expand and to resonate in your cheeks as well as in your sinuses. Again with a normal breath, easily, say "Hi!" resonating in your sinuses, cheeks, and upper chest. Then, without adding any more effort, add the belly as a resonator. Notice how the volume of your "Hi!" changes as you include more of your body's resonant chambers.*

As children, we imitate styles of speaking and learn patterns of language and sound. Like the hands and feet, our voice provides refined connection for communicating and for manipulating our environment for survival and interaction. As adults, we have choice about how we use our voice. We can affect what our voice sounds like and how clearly it reflects our thoughts and feelings: how we speak and what we choose to say. Use of language and use of tone of voice have tremendous effect. As we are free to relax and breathe deeply, the tones that emerge will reflect who we are.

Sound gives expression to the body. Every molecule in the universe is vibrating. When we produce sound in the body, we are making vibration audible. As we have seen, we use the force of our breath to vibrate the

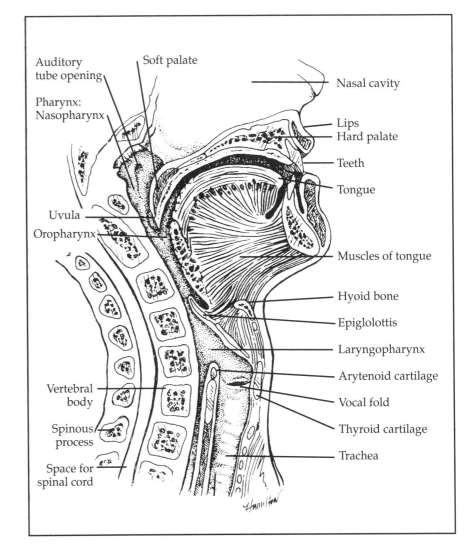

**Sagittal section
showing vocal mechanisms**

vocal cords with higher frequencies resonating in the small chambers such as the sinuses in the skull and the lower frequencies in areas such as the pelvis and belly. Sound literally vibrates the body. Different tissues respond according to their density and elasticity. By producing sounds with our body: singing, humming, chanting, toning, and playing instruments, we are vibrating ourselves as well as others. As we change our pitch, we give ourself an inner massage.

Singing clears the pathways in the body. We all know how much better we feel when we sing. (Why did we stop?) By vibrating the tissues, muscles relax, fluids flow, nerves calm. Try this in your own body: find a tense area, such as the neck muscles. Place your hands on the area to focus your attention (on the body part) and your intention (on vibration). Make a sound that vibrates into your hand through the neck tissues. Feel the muscle tension dissolve. Bring your awareness to any tension in your skull. Place your hands on the cranial bones: make a high tone and feel the vibration with your fingers through the bones of the skull. This brings awareness to the area and helps you to feel the change. If your lower back hurts, or you have difficulty with the digestive organs, place your hands on the particular area of the belly or back. Use a low tone to vibrate the tissues. Breathe deeply and naturally as you work. No need to judge the quality of the sound, just feel its vibration. Observe what happens.

Sound carries emotion. We have all experienced love songs stimulating our hearts, a cry freezing us in terror, a tone of voice bringing tears to our eyes. Emotions ride on sound and can be released, transformed, or communicated. Sometimes we close our throats to try to hold back emotion, the familiar lump in the throat or tightness in a voice. By listening to our sounds, we can come to know our feelings. Sound moves us. Music and speaking have been used for centuries to heal, to entice, to calm, to stimulate. Each tone, each instrument, plays the body in a different way. Rhythms, melodies, instrumentation, words and intonations all have varying effects. Both sound and silence affect our body functioning.

Sound exists outside the boundaries of known forms such as music and language. As we allow unmonitored sounds to emerge from the body and hear their qualities, we become familiar with new spectrums of experience. Free sound is unpredictable, and we have no idea where it comes from or will go; it creates its own form, and we get to follow. As we listen to music from other cultures, we open our awareness. Eskimo throat singing, Tibetan chanting, the Chinese erhu offer new possibilities. As we extend our work with Authentic Movement to include sound, we find a rich way to listen to the body and to hear ourselves.

*Description by Susan Borg, a bodyworker and Resonant Kinesiologist in Burlington, Vermont.

Yacalecuhtli, the masked god of commerce, showing the scroll-like symbol for singing. From a mural at Teotihuacán, Mexico.

WORKING WITH SOUND

As we explore sound and movement we can "play" our bodies by producing sounds with our voices. We can tone our organs, and vibrate into tight muscles for increased circulation and release. Clapping and stomping of feet activate the blood and stimulate circulation. We can "be played" by music on radios, at concerts, and sounds on the street. We can allow sounds to emerge which are not permitted in daily activity, but which the body has stored and needs to express. We can also take time to listen to the sound within silence.

Spontaneous sound is a way of developing a dialogue between

what we make happen and what we allow to happen.

There is plenty going on inside of us if we choose to listen.

Sounds, pulses and rhythms connect us fluidly

to the rest of the world.

Kokopelli, the ancient Anasazi god of fertility, carried seeds and played his flute to warm their germination. He is often shown as a humpback, wearing antennae or with exaggerated genitals in hieroglyphic drawings on cliff walls found throughout the Southwest U.S.

154

Moving with sound
20 minutes

Lying on the floor in constructive rest:
○ Observe your breathing. Allow your jaw to relax open.
○ Begin following any impulse in your body for movement. It might be a leg, it might be an arm; any body part that wants to move. Whatever feels good is what you want to do.
○ See if there is any sound that wants to come with your movement. A sigh, a yawn, a growl, a tone. If words or sounds come, let them accompany your movement, but keep your focus on the sensations in the body. Continue moving, inviting sound.
○ Continue for ten minutes. Bring your work to a close. Roll to seated and open your eyes.

Vibrating the body, constructive rest
15 minutes

Lying in constructive rest, your hands on your ribs:
○ Feel your breathing.
○ Make an open sound, like an "ah" that vibrates your ribs.
○ Continue to sound into this area easily on the out breaths.
○ Move your hands to your belly. Find a tone that vibrates your belly. Feel the vibrations in your hands.
○ Place your hands on your solar plexus, below the ribs.
Find a tone that vibrates the solar plexus.
○ Place your hands on your sternum. Tone into your sternum.
Feel the vibration in your hands as well as your sternum.
○ Place your hands on your neck. Vibrate sound in your neck and hands.
○ Place your hands on your TMJ joint of the jaw. Tone into your jaw.
○ Place your hands on your forehead. Sound into your forehead. Continue until the whole skull vibrates.
○ Place your hands on the top of the skull. Vibrate the top of the skull with sound.
○ Reverse, from head to pelvis. See how low a sound you can make to vibrate in the pelvis and hip sockets.
○ Roll to your side and come to standing.

Vibrating the body with sound, standing*
5 minutes

Standing: Hang forward from your waist, head and arms relaxed, knees bent:

❍ Make as low an "ah" sound as you can, vibrating in your feet and pelvis. Bounce your legs a few times to help drop the sound down through the body. Image a ball bouncing on the pelvic floor with each sound, like on a trampoline.

❍ Begin rolling your sound up the front of your spine from your lowest tone up sequentially through your highest tone. As you do this, roll up to plumb line. Jump in the air as you bounce the high tone off of the top of the skull. Let the arms fly up as you jump.

❍ Repeat the high sound and the jumping a few times to feel the vibration. Image it going right out the ceiling, like an elevator going up.

❍ Reverse. Standing in plumb line, send a tone into the skull with a high sound. Part by part, relax the body weights back down to the earth as the sound descends. Image an elevator going down into the basement. Release the neck and jaw as you tone. Bounce a few times, bringing the sound as low as you can.

❍ Repeat several times. Observe how you feel.

*Kristin Linklater has many useful exercises for sound and movement in her book, *Freeing the Natural Voice.*

PERSONAL PROJECT

There are many ways of looking at the body. What catches your interest is yours to follow. Reflect on your bodystory, and identify an area of your body which is of particular interest to you. It might be an area which has been injured, or which you ignore. It could be a body part which you particularly like, or which stimulates images or ideas. A body scan might help you select.

Information: Identify the bones and joints involved in the area of your choice. Research other aspects which draw your attention: muscles, connective tissue, nerves, fluids, organs, endocrines. Be specific.

Movement: From all the exercises that we have studied, select what you want to do to affect change or bring awareness to this area of your body. Be consistent with the work over a period of time.

Stories: Write stories about your selected body area.

Give yourself two weeks to work on your project. Follow your interest. ❖

Endings as Beginnings

Teaching the creative process, I often speak of its cyclic nature: once the creative impulse is acknowledged, we find a medium for expression, identify its context and content, let it change and transform, and then, quite often, a finished form emerges almost identical to the initial impulse. But the entire process is necessary to arrive at the end, which is like the beginning. It reminds me of one of my favorite lines in Edward Albee's **Zoo Story**: *"Sometimes it is necessary to go a very long distance out of your way, to come back a short distance correctly."*

❖

As I was completing this book, my stories of endings came forward. The process had begun many years before. I gave the draft to readers, a tall stack of pages organized by two silver rings. One response, "The rings are important. As I turned the pages, the back became the front. When I arrived at the end, I was at the beginning. That seemed right. Learning about the body is an ongoing process. How do you keep that present in book form?"

❖

I talked to a friend about endings. She said that after the death of her husband she put together every jigsaw puzzle in the house. When she was done, the pieces were in place to begin again. Another friend said that each time we begin we start from our base. If we start in the middle, we can only do what we know. If we start from the beginning, our possibilities are endless.

**Photograph: Bill Arnold
"Studio: Theatre School, Amsterdam"**

157

Drawing: Hally Sheely (age 4)

Following your movement

Lying in constructive rest, eyes closed:
○ Do a full body scan.
○ Then bring your awareness to your particular body area. Follow any movement that feels good. You may be drawn to an image, a sensation, an idea, an emotion.
○ Listen to whatever comes. Allow yourself to move and let your body be your guide.

Teach-a-friend

○ Work with someone for three days, a minimum of 30 - 45 minute sessions. Select "To do" exercises from the text which were particularly useful and clear to you.
○ Work in a neutral, private environment. Minimize talk time – focus on hands-on work.
Note: Sometimes it is useful to repeat part of an exercise from one day before introducing new work. For example, on the second day, begin by "Holding the head," and then continue to "Tracing the ribs," or repeat "Hip circles" before going on to "Foot massage." Ask your friend for feedback at the end of the three sessions: What did they feel? What was most useful to them? Do they have any suggestions or questions?

Teach-a-friend paper

○ Ask your friend to write a brief description of her/his experience – hand written.
○ Write a three-five page paper describing in detail your own experience: Which exercises you used, the length of time for each, what kind of space you worked in, the types of touch which were most useful and what you learned.
○ Evaluate your experience.
○ Read your friend's comments.

BIBLIOGRAPHY

- Adler, Janet. "Who is the Witness? – A Description of Authentic Movement." *Contact Quarterly*, Vol.12:1. 1987. Interviews.
  _____. *Still Looking* (film). U. of California Resource Center, Berkeley.
- Alexander, F. Matthias. *The Alexander Technique*. Carol Publishing Group, NY, 1967, 1968.
- Attenborough, David. *Life on Earth, A Natural History*. Little, Brown, and Company, Boston, Toronto, 1979.
- Bartenieff, I. and Lewis, D. *Body Movement: Coping with the Environment*. Gordon and Breach, NY, 1980.
- Cohen, Bonnie. Articles, *Contact Quarterly*: CQ Reprint No. 1: "The Neuroendocrine System," "Sensing, Feeling and Action." Vol. 9: 2 incl."Perceiving in Action – Interview on the Developmental Process." Vol. 10:2 incl. "The Mechanics of Vocal Expression." Vol.12:3 incl. "The Action in Perceiving." Vol. 13:3 incl. "The Dancer's Warm-up." Vol. 14:2 incl. "The Alphabet of Movement, Part 1." Vol. 14:3 incl. "The Alphabet of Movement, Part 2." Vol. 16:2 incl. "The Dynamics of Flow – The Fluid System of the Body." Interviews 1989, 1990.
- Cowan, W. M. "The Development of the Brain." *Scientific American*, Sept. 1979.
- Dicara, L.V. "Learning in the Autonomic Nervous System." *Scientific American*, Jan. 1970.
- Dowd, Irene. "Taking Root To Fly, Ten Articles on Functional Anatomy." Contact Collaborations, Northampton, Ma., 1981.
- Dychtwald, Ken. *Bodymind*. Pantheon Books, NY, 1977.
- Elson, Lawrence; Kapit, Wynn. *The Anatomy Coloring Book*. Harper and Row, NY, 1977.
- Feldenkrais, Moshe. *Awareness Through Movement*. Harper and Row, NY, 1990.
- Fitt, Sally Sevey. *Dance Kinesiology*. Schirmer Books, Macmillan, NY, 1988.
- Gorman, David. *The Body Moveable*. Ampersand Printing Company, Canada, 1981.
- H'Doubler, Margaret. *Dance: A Creative Art Experience*. University of Wisconsin Press, Madison, 1957.
- Huang, Chungliang Al. *Embrace Tiger, Return to Mountain; The Essence of Tai Chi*. Celestial Arts, Berkeley, 1973, 1987.
- Huxley, H.E. "The Contraction of Muscle." *Scientific American*, Nov. 1958.
- Jacob, Francone, Lossow. *Structure and Function In Man*. W.B. Saunders Company, Philadelphia, 1965, 1982.
- Johnson, Don. *The Protean Body, A Rolfer's View of Human Flexibility*. Harper Colophon Books, Harper and Row, 1977.
- Juhan, Deanne. *Job's Body, A Handbook for Bodywork*. Station Hill Press, Barrytown, NY, 1987.
- Jung, C.G. *Man and His Symbols*. Anchor Press, NY, 1964.
  _____. *Psyche and Symbol*. Anchor Press, NY, 1958.
- Katz, B. "How Cells Communicate." *Scientific American*, September, 1961.
- Lewin, Roger. *Thread of Life, The Smithsonian Looks at Evolution*. Smithsonian Books, Washington, D.C., 1982.
- Leibowitz, Judith and Connington, Bill. *The Alexander Technique*. Harper and Row, NY, 1990.
- Linklater, Kristin. *Freeing the Natural Voice*. Drama Book Specialists, NY, 1976.
- Montagu A. *Touching: The Human Significance of Skin*. Harper and Row, NY, 1971.

NOTES

❖

Special thanks to my parents, my students, my friends and my colleagues whose lives are the heart and substance of these writings; and to Middlebury College as an institution which encourages the creative work of its faculty.

❖

Particular acknowledgment goes to: Rosalyn Driscoll and Harriet Brickman, who helped to shape and edit the text; Kristina Madsen, Anne Woodhull and Kiki Smith, who listened to my stories; Tom Root and Susan Borg, who added their resources in biology and voice; Sandra Jamrog, who assisted in the description of birthing; Karen Murley, who brought the work into visual form through graphic design; and Erik Borg, whose photographs illustrate the text.

- Napier, John. "The Antiquity of Human Walking." *Scientific American*, April, l961.
- Nilsson, Lenart. *Behold Man.* Little, Brown and Co., Boston, 1973.
- Rolf, Ida. *Rolfing, The Integration of Human Structures.* Healing Arts Press, Rochester, Vt. 1977/1989.
- Rose, Kenneth Jon. *The Body In Time.* Wiley Science Editions, NY, 1988.
- Schlossberg, Leon. Zuidema, George D. *The Johns Hopkins Atlas of Human Anatomy.* The Johns Hopkins University Press, 1972/1986.
- Snyder, S. H. "The Molecular Basis of Communication Between Cells." *Scientific American*, Oct. 1985.
- Steinman, Louise. *The Knowing Body.* Shambala Press, Boston, 1986.
- Swiegard, Lulu. *Human Movement Potential, It's Ideokinetic Facilitation.* Harper and Row, NY, l974.
- Todd, Mabel. *The Thinking Body, A Study of the Balancing Forces of Dynamic Man.* Dance Horizons Inc., Brooklyn, l939.
- Tortora, G. and Anagnostakos, N. *Principles of Anatomy and Physiology.* Fifth Edition. Harper and Row, NY, 1987.
- Trager, Milton. *Mentastics, Movement As A Way To Agelessness.* Station Hill Press, Barrytown, NY, 1987.
- Upledger, John. Vredovoogd, Jon. *Craniosacral Therapy.* Eastland Press, Seattle, 1983.
- Vannini, V. and Pagliani, G., eds. *The Color Atlas of Human Anatomy.* Translated and revised by Dr. Richard T. Jelly, Harmony Books, NY, 1980.
- Warfel, John H. *The Head, Neck and Trunk.* Lea and Ferbinger, Philadelphia, 1973.
- Wilson, John M. *A Natural Philosophy of Movement Styles for Theatre Performers.* (Doctoral dissertation; the University of Wisconsin-Madison: 1973), pp. 122-129.

  _____. "Kinesiology and the Art of Centering" in *Kinesiology for Dance!* Newsletter. UCLA Department of Dance (nos. 3, 4, 5, 6; Aug. 1977 – Jan. 1979), Martin Tracy, Ed.

  _____. "Dance Takes a New Partner," *The University of Arizona Arts*, Vol. 2, No. 2; (Summer 1981-1982), pp. 8-14.

  _____. "Movement Styles in Dance and Visual Arts: Viewing Through the Lens of the Arthrometric Model," Congress on Research in Dance, annual conference. University of Iowa: Nov. 1991.

VISUAL IMAGERY

- Anatomical Drawings: William P. Hamilton, medical illustrator.
  Pages 40, 62, 66, 72, 73, 84, 88, 93, 98, 106, 119, 128, 142, 152.
- Photographs of bones: Erik Borg.
  Pages v, 40, 43, 44, 50, 57, 66, 67, 72, 73, 75, 84, 98, 102, 106.
- Photographs of hands-on work (Caryn McHose, Andrea Olsen):
  Erik Borg. Pages 4, 13, 21, 45, 46, 70, 107, 136, 137, 138.
- Computer generated illustrations: Karen Murley.
  Pages 14, 16, 32, 35, 47, 50, 56, 69, 76, 89, 114.
- Art images: Bill Arnold – 22, 30, 48, 101, 157; Erik Borg – 15, 105;
  Harriet Brickman – 38, 83, 112, 140; Jim Butler – 42, 75, 118;
  Rosalyn Driscoll – 18, 61; Robert Ferris – 11, 130, 146;
  John Huddelston – 126; Kristen Kagan – 55, 105;
  Kristina Madsen – 6, 71, 92; Michael Singer – 34, 78, 87, 96;
  Gordon Thorne – 26, 65, 150; Whitney Sander – 28.

ART INDEX

MULTICULTURAL IMAGES

* Line drawings by Karen Murley, adapted from the source

SUBJECT INDEX

ANDREA OLSEN, author, is an Associate Professor of Dance at Middlebury College in Vermont. She also directed the dance program at Mount Holyoke College, and the Dance Gallery Schools in Northampton, Massachusetts, and Seattle, Washington. An MFA graduate of the University of Utah in Dance, she has taught anatomy and kinesiology since 1972 in workshops and colleges. She has worked with colleagues Bonnie Bainbridge Cohen in experiential anatomy, and Janet Adler in Authentic Movement since 1979, while developing an approach to hands-on teaching of anatomy for college students with Caryn McHose. As a dancer, she has choreographed over fifty works and toured internationally for twenty years with Dance Gallery and the Dance Company of Middlebury. Most recently, she has completed solo performing tours in China, Japan, Hong Kong, New Zealand, Holland and Greece presenting work developed in collaboration with musician David Darling, and traveled in Kenya and Tanzania. Thornes Market in Northampton, Massachusetts is her creative home, and she continues to teach and perform there on a regular basis. ❖

CARYN McHOSE, collaborator, received her BFA in Art and Dance from the University of Connecticut and studied with Bonnie Bainbridge Cohen in Amherst, Massachusetts. She spent six years as Lecturer at Middlebury College in Vermont teaching dance and exploring the practical application of anatomy and kinesiology for dance, theatre, and athletic performance. She also toured and taught for the Vermont Council on the Arts, including workshops for children and for senior citizens. Her work in anatomy and dance has been influenced by vocal study with Susan Borg and early dance training with Betty Jane Dittmar. She has had a private practice in bodywork in Vermont and Maine since 1979. ❖